The Circuit Designer's Companion

The Circuit Designer's Companion

Tim Williams

Butterworth-Heinemann Ltd
Halley Court, Jordan Hill, Oxford OX2 8EJ

 PART OF REED INTERNATIONAL BOOKS

OXFORD LONDON GUILDFORD BOSTON
MUNICH NEW DELHI SINGAPORE SYDNEY
TOKYO TORONTO WELLINGTON

First published 1991

British Library Cataloguing in Publication Data
Williams, Tim
   The circuit designer's companion
   1. Circuits
   I. Title
   621.38153

Library of Congress Cataloguing in Publication Data
Williams, Tim
      The circuit designer's companion/Tim Williams.
         p.   cm.
      Includes bibliographical references and index.
      ISBN 0 – 7506 – 1142 – 1
      1. Electronic circuit design.   I. Title.
   TK7867.W55   1991
   621.381'5 – – dc20   90 – 26104

ISBN 0 7506 1142 1

Printed and bound in Great Britain by
Billing and Sons Ltd, Worcester

Contents

Chapter 3
Passive Components 63

Introduction

Electronic circuit design can be divided into two areas: the first consists in designing a circuit that will fulfil its specified function, sometimes, under laboratory conditions; the second consists in designing the same circuit so that every production model of it will fulfil its specified function, and no other undesired and unspecified function, always, in the field, reliably over its lifetime. When related to circuit design skills, these two areas coincide remarkably well with what engineers are taught at college – basic circuit theory, Ohm's Law, Thévenin, Kirchoff, Norton, Maxwell and so on – and what they learn on the job – that there is no such thing as the ideal component, that printed circuits are more than just a collection of tracks, and that electrons have an unfortunate habit of never doing exactly what they're told.

This book has been written with the intention of bringing together and tying up some of the loose ends of analogue and digital circuit design, those parts that are never mentioned in the textbooks and rarely admitted elsewhere. In other words, it relates to the second of the above areas.

Its genesis came with the growing frustration experienced as a senior design engineer, attempting to recruit people for junior engineer positions in companies whose foundations rested on analogue design excellence. Increasingly, it became clear that the people I and my colleagues were interviewing had only the sketchiest of training in electronic circuit design, despite offering apparently sound degree-level academic qualifications. Many of them were more than capable of hooking together a microprocessor and a few large-scale functional block peripherals, but were floored by simple questions such as the nature of the p-n junction or how to go about resistor tolerancing. It seems that this experience is by no means uncommon in other parts of the industry.

The colleges and universities can hardly be blamed for putting the emphasis in their courses on the skills needed to cope with digital electronics, which is after all becoming more and more pervasive. If they are failing industry, then surely it is industry's job to tell them and to help put matters right. Unfortunately it is not so easy. A 1989 report from Imperial College, London, found that few students were attracted to analogue design, citing inadequate teaching and textbooks as well as the subject being found "more difficult". Also, teaching institutions are under continuous pressure to broaden their curriculum, to produce more "well-rounded" engineers, and this has to be at the expense of greater in-depth coverage of the fundamental disciplines.

Nevertheless, the real world is obstinately analogue and will remain so. There is a disturbing tendency to treat analogue and digital design as two entirely separate disciplines, which does not result in good training for either. Digital circuits are in reality only over-driven analogue ones, and anybody who has a good understanding of analogue principles is well placed to analyse the more obscure behaviour of logic devices. Even apparently simple digital circuits need some grasp of their analogue interactions to be designed properly, as chapter 6 of this book shows. But also, any product which interacts with the outside world via typical transducers must contain at least some analogue circuits for signal conditioning and the supply of power. Indeed, some products are still best realised as all-analogue circuits. Jim Williams, a well-known American linear circuit designer (who bears no relation to this author), put it succinctly when he said "wonderful things are going on in the forgotten land between ONE and ZERO. This is Real Electronics".

Because analogue design appears to be getting less popular, those people who do have such skills will become more sought-after in the years ahead. This book is meant to be a tool for any aspiring designer who wishes to develop these skills. It assumes at least a background in electronics design; you will not find in here more than a minimum of basic circuit theory. Neither will you find recipes for standard circuits, as there are many other excellent books which cover those areas. Instead, there is a serious treatment of those topics which are "more difficult" than building-block electronics: grounding, temperature effects, EMC, component sourcing and characteristics, the imperfections of devices, and how to design so that someone else can make the product.

I hope the book will be as useful to the experienced designer who wishes to broaden his or her background as it will to the neophyte fresh from college who faces a first job in industry with trepidation and excitement. The traditional way of gaining experience is to learn on-the-job through peer contact, and this book is meant to enhance rather than supplant that route. It is offered to those who want their circuits to stand a greater chance of working first time every time, and a lesser chance of being completely re-designed after six months. It does not claim to be conclusive or complete. Electronic design, analogue or digital, remains a personal art, and all designers have their own favourite tricks and their own dislikes. Rather, it aims to stimulate and encourage the quest for excellence in circuit design.

I must here acknowledge a debt to the many colleagues over the years who have helped me towards an understanding of circuit design, and who have contributed towards this book, some without knowing it: in particular Tim Price, Bruce Piggott and Trevor Forrest. Also to Joyce, who has patiently endured the many brainstorms that the writing of it produced in her partner.

Tim Williams
October 1990

Chapter 1

Grounding and wiring

1.1   Grounding

A fundamental property of any electronic or electrical circuit is that the voltages present within it are referenced to a common point, conventionally called the ground. (This term is derived from electrical engineering practice, when the reference point is often taken to a copper spike literally driven into the ground.) This point may also be a connection point for the power to the circuit, and it is then called the 0V (nought-volt) rail, and ground and 0V are frequently (and confusingly) synonymous. Then, when we talk about a five-volt supply or a minus-twelve-volt supply or a two-and-a-half-volt reference, each of these is referred to the 0V rail.

At the same time, ground is *not* the same as 0V. A ground wire connects equipment to earth for safety reasons, and does not carry a current in normal operation. However, in this chapter the word "grounding" will be used in its usual sense, to include both safety earths and signal and power return paths.

Perhaps the greatest single cause of problems in electronic circuits is that 0V and ground are taken for granted. The fact is that in a working circuit there can only ever be one point which is truly at 0V; the concept of a "0V rail" is in fact a contradiction in terms. This is because any practical conductor has a finite non-zero resistance and inductance, and Ohm's Law tells us that a current flowing through anything other than a zero impedance will develop a voltage across it. A working circuit will have current flowing through those conductors that are designated as the 0V rail and therefore, if any one point of the rail is actually at 0V (say, the power supply connection) the rest of the rail will *not* be at 0V. This is illustrated in Figure 1.1.

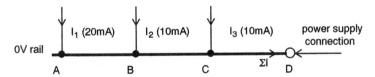

Assume the 0V conductor has a resistance of 10mΩ/inch and that points A, B, C and D are each one inch apart. The voltages at points A, B and C referred to D are

$$V_C = (I_1 + I_2 + I_3) \cdot 10m\Omega \qquad = 400\mu V$$

$$V_B = V_C + (I_1 + I_2) \cdot 10m\Omega \qquad = 700\mu V$$

$$V_A = V_B + (I_1) \cdot 10m\Omega \qquad = 900\mu V$$

Figure 1.1 Voltages along the 0V rail

Now, after such a trenchant introduction, you might be tempted to say well, there are millions of electronic circuits in existence, they must all have 0V rails, they seem to work well enough, so what's the problem? Most of the time there is no problem. The impedance of the 0V conductor is in the region of milliohms, the current levels are milliamps, and the resulting few hundred microvolts drop doesn't offend the circuit at all. 0V plus 500μV is close enough to 0V for nobody to worry.

The difficulty with this answer is that it is then easy to forget about the 0V rail and assume that it is 0V under all conditions, and subsequently be surprised when a circuit oscillates or otherwise doesn't work. Those conditions where trouble is likely to arise are

- where current flows are measured in amps rather than milli- or microamps
- where the 0V conductor impedance is measured in ohms rather than milliohms
- where the resultant voltage drop, whatever its value, is of a magnitude or in such a configuration as to affect the circuit operation

When to consider grounding

One of the attributes of a good circuit designer is to know when these conditions need to be carefully considered and when they may be safely ignored. A frequent complication is that you as circuit designer may not be responsible for the circuit's layout, which is handed over to a layout draughtsman (who may in turn delegate many routing decisions to the computer-aided design software). Grounding is always sensitive to layout, whether of discrete wiring or of printed circuits, and the designer must have some knowledge of and control over this if the design is not to be compromised.

The trick is always to be sure that you know where ground return currents are flowing, and what their consequences will be; or, if this is too complicated, to make sure that wherever they flow, the consequences will be minimal. Although the above comments are aimed at 0V and ground connections, because they are the ones most taken for granted, the nature of the problem is universal and applies to any conductor through which current flows. The power supply rail (or rails) is another special case where conductor impedance can create difficulties.

1.1.1   Grounding within one unit

In this context, "unit" can refer to a single circuit board or a group of boards and other wiring connected together within an enclosure such that you can identify a "local" ground point, for instance the point of entry of the mains earth. An example might be as shown in Figure 1.2. Let us say that printed circuit board (PCB) 1 contains input signal conditioning circuitry, PCB2 contains a microprocessor for signal processing and PCB3 contains high-current output drivers, such as for relays and for lamps. You may not place all these functions on separate boards, but the principles are easier to outline and understand if they are considered separately. The power supply unit (PSU) provides a low-voltage supply for the first two boards, and a higher-power supply for the output board. This is a fairly common system layout and Figure 1.2 will serve as a starting point to illustrate good and bad practice.

1.1.2   Chassis ground

First of all, note that connections are only made to the metal chassis or enclosure at one

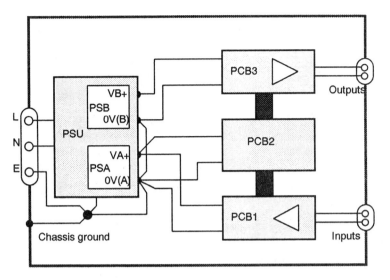

Figure 1.2 Typical intra-unit wiring scheme

point. All wires that need to come to the chassis are brought to this point, which should be a metal stud dedicated to the purpose. Such connections are the mains safety earth (about which more later), the 0V power rail, and any possible screening and filtering connections that may be required in the power supply itself, such as an electrostatic screen in the transformer. (The topic of power supply design is itself dealt with in much greater detail in chapter 6).

The purpose of a single-point chassis ground is to prevent circulating currents in the chassis.[†] If multiple ground points are used, even if there is another return path for the current to take, a proportion of it will flow in the chassis (Figure 1.3). Such currents are very hard to predict and may be affected by changes in construction, so that they can give quite unexpected and annoying effects: it is not unknown for hours to be devoted to tracking down an oscillation or interference problem, only to find that it disappears when an inoffensive-looking screw is tightened against the chassis plate. Another good reason for avoiding chassis currents is that joints in the chassis are affected by corrosion, so that the unit performance may degrade with time, and they are affected by surface oxidation of the chassis material.

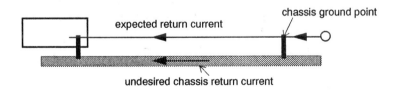

Figure 1.3 Return current paths with multiple ground points

† Note that, when rf shielding and/or a low-inductance ground is required, multiple ground points may be essential. This is covered in chapter 8.

1.1.3 The conductivity of aluminium

Aluminium is used throughout the electronics industry as a light, strong and highly conductive chassis material – only silver, copper and gold have a higher conductivity. You would expect an aluminium chassis to exhibit a decently low bulk resistance, and so it does, and is very suitable as a conductive ground as a result. Unfortunately, another property of aluminium (which is useful in other contexts) is that it oxidises very readily on its surface, to the extent that all real-life samples of aluminium are covered by a thin surface film of aluminium oxide (Al_2O_3). Aluminium oxide is an insulator. In fact, it is such a good insulator that anodised aluminium, on which a thick coating of oxide is deliberately grown by chemical treatment, is used for insulating washers on heatsinks.

The practical consequence of this quality of aluminium oxide is that the contact resistance of two sheets of aluminium jointed together is unpredictably high. Actual electrical contact will only be made where the oxide film is breached. Therefore, whenever you want to maintain continuity through a chassis made of separate pieces of aluminium, you must ensure that the plates are tightly bonded together, preferably with welding or by fixings which incorporate shakeproof serrated washers to actively dig into the surface. The same applies to ground connection points. The best connection (since aluminium cannot easily be soldered) is a force-fit or welded stud (Figure 1.4), but if this is not available then a shakeproof serrated washer should be used underneath the nut which is in contact with the aluminium.

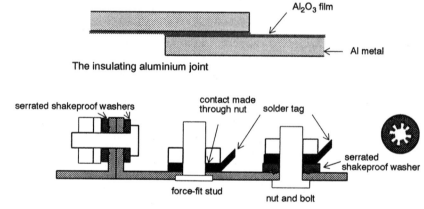

Figure 1.4 Electrical connections to aluminium

Other materials

The other common chassis material is cadmium-plated steel, which does not suffer from the oxidation problem. Mild steel has about three times the bulk resistance of aluminium so does not make such a good conductor, but it has better magnetic shielding properties and it is cheaper. Other metals, particularly silver-plated copper, can be used where the ultimate in conductivity is needed and cost is secondary, as in rf circuits. The advantage of silver oxide (which forms on the silver-plated surface) is that it is conductive and can be soldered through easily.

1.1.4 Ground loops

Another reason for single-point chassis connection is that circulating chassis currents,

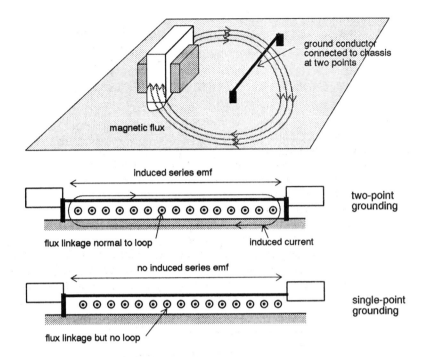

ground conductor connected to chassis at two points

magnetic flux

Induced series emf

flux linkage normal to loop   induced current

two-point grounding

no induced series emf

flux linkage but no loop

single-point grounding

Figure 1.5 The ground loop

when combined with other ground wiring, produce the so-called "ground loop", which is a fruitful source of low-frequency magnetically-induced interference. A magnetic field can only induce a current to flow within a closed loop circuit. Magnetic fields are common around power transformers – not only the conventional 50Hz mains type (60Hz in the US), but also high-frequency switching transformers and inductors in switched-mode power supplies – and also other electromagnetic devices: contactors, solenoids and fans. Extraneous magnetic fields may also be present. The mechanism of ground-loop induction is shown in Figure 1.5.

Lenz's law tells us that the e.m.f. induced in the loop is

$$V = -10^{-8} \cdot A \cdot n \cdot dB/dt$$

where   A is the area of the loop in cm^2
B is the flux density normal to it, in Gauss, assuming a uniform field
n is the number of turns (n = 1 for a single-turn loop)

As an example, take a 10 Gauss 50Hz field as might be found near a reasonable-sized mains transformer, contactor or motor, acting through a $10cm^2$ loop that would be created by running a conductor 1cm above a chassis for 10cm and grounding it at both ends. The induced emf is given by

$$V = -10^{-8} \cdot 10 \cdot d/dt(10 \cdot \sin 2\pi \cdot 50 \cdot t)$$

$$= -10^{-8} \cdot 10 \cdot 1000\pi \cdot \cos \omega t$$

$$= \underline{\textbf{314\mu V peak}}$$

Magnetic field induction is a low-frequency phenomenon (unless you happen to be very close to a high-power radio transmitter) and you can see from this example that in most circumstances the induced voltages are low. But in low-level applications, particularly audio and precision instrumentation, they are far from insignificant. If the input circuit includes a ground loop, the interference voltage is injected directly in series with the wanted signal and cannot then be separated from it. The cures are:

- open the loop by grounding only at one point
- reduce the area of the loop (A in the equation above) by routing the offending wire(s) right next to the ground plane or chassis, or shortening it
- reduce the flux normal to the loop by repositioning or reorienting the loop or the interfering source
- reduce the interfering source, for instance by using a toroidal transformer

1.1.5   Power supply returns

You will note from Figure 1.2 that the output power supply 0V connection (0V(B)) has been shown separately from 0V(A), and linked only at the power supply itself. What happens if, say for reasons of economy in wiring, you don't follow this practice but instead link the 0V rails together at PCB3 and PCB2, as shown in Figure 1.6?

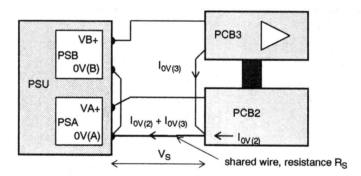

Figure 1.6 Common power supply return

The supply return currents I_{0V} from both PSB/PCB3 and PSA/PCB2 now share the same length of wire (or track, in a single-pcb system). This wire has a certain non-zero impedance, say for dc purposes it is R_S. In the original circuit this was only carrying $I_{0V(2)}$ and so the voltage developed across it was

$$V_S \quad = \quad R_S \cdot I_{0V(2)}$$

but, in the economy circuit,

$$V_S \quad = \quad R_S \cdot (I_{0V(2)} + I_{0V(3)})$$

This voltage is in series with the supply voltages to both boards and hence effectively subtracts from them.

Putting some typical numbers into the equations,

$I_{0V(3)}$ = 1.2A   with a VB+ of 24V because it is a high-power output board,

$I_{0V(2)}$ = 50mA   with a VA+ of 5V because it is a microprocessor board with some LS-TTL logic on it.

Now assume that, for various reasons, the power supply is some distance remote from the boards and you have without thinking connected it with six feet of 7/0.2mm equipment wire, which will have a room temperature resistance of about 0.2Ω. The voltage V_S will be

$$V_S = 0.2 \cdot (1.2 + 0.05) = 0.25V$$

which will drop the supply voltage at PCB2 to 4.75V, the lower limit of operation for LS-TTL, *before* allowing for supply voltage tolerances and other voltage drops. One wrong wiring connection can make your circuit operation borderline! Of course, the 0.25V is also subtracted from the 24V supply, but a reduction of about 1% on this supply is unlikely to affect operation.

Varying loads

If the 1.2A load on PCB3 is varying - say several high-current relays may be switched at different times, ranging from all off to all on - then the V_S drop at PCB2 would also vary. This is very often worse than a static voltage drop because it introduces noise on the 0V line. The effects of this include unreliable processor operation, variable set threshold voltage levels and odd feedback effects such as chattering relays or, in audio circuits, low-frequency "motor-boating" oscillation.

For comparison, look at the same figures but applied to Figure 1.2, with separate 0V return wires. Now there are two voltage drops to consider: $V_{S(A)}$ for the 5V supply and $V_{S(B)}$ for the 24V supply. $V_{S(B)}$ is 1.2A times 0.2Ω, substantially the same (0.24V) as before, but it is only subtracted from the 24V supply. $V_{S(A)}$ is now 50mA times 0.2Ω or 10mV, which is the only 0V drop on the 5V supply to PCB2 and is negligible.

The rule is: always separate power supply returns so that load currents for each supply flow in separate conductors (Figure 1.7).

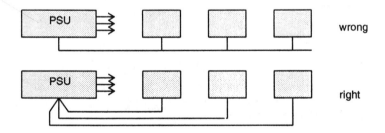

Figure 1.7 Ways to connect power supply return

Note that this rule is easiest to apply if different power supplies have different 0V connections (as in Figure 1.2) but should also be applied if a common 0V is used, as shown above. The extra investment in wiring is just about always worth it for peace of mind!

Power rail feed

The rule also applies to the power rail feed as well as to its return, and in fact to any connection where current is being shared between several circuits. Say the high-power load on PCB3 was also being fed from the +5V supply VA+, then the preferred method of connection is two separate feeds (Figure 1.8).

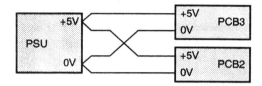

Figure 1.8 Separate power supply rail feeds

   The reasons are the same as for the 0V return: with a single feed wire, a common voltage drop appears in series with the supply voltage, injected this time in the supply rail rather than the 0V rail. Fault symptoms are similar. Of course, the example above is somewhat artificial in that you would normally use a rather more suitable size of wire for the current expected. High currents flowing through long wires demand a low-resistance and hence thick conductor. If you are expecting a significant voltage drop then you will take the trouble to calculate it for a given wire diameter, length and current. See Table 1.2 for a guide to the current-carrying abilities of common wires. The point of the previous examples is that voltage drops have a habit of cropping up when you are *not* expecting them.

Conductor impedance

Note that the previous examples, and those on the next few pages, tacitly assume for simplicity that the wire impedance is resistive only. In fact, real wire has inductance as well as resistance and this comes into effect as soon as the wire is carrying ac, increasing in significance as the frequency is raised. A one-metre length of 16/0.2 equipment wire has a resistance of 38mΩ and an inductance of 1.5μH. At 4A dc the voltage drop across it will be 152mV. An ac current with a rate of change of 4A/μs will generate 6V across it. Note the difference! The later discussion of wire types includes a closer look at inductance.

1.1.6   Input signal ground

Figure 1.2 shows the input signal connections being taken directly to PCB1 and not grounded outside of the pcb. To expand on this, the preferred scheme for two-wire single-ended input connections is to take the ground return directly to the reference point of the input amplifier: see Figure 1.9(a).

   The reference point on a single-ended input is not always easy to find: look for the point from which the input voltage must be developed in order for the amplifier gain to act on it alone. In this way, no extra signals are introduced in series with the wanted signal by means of a common impedance. In each of the examples in Figure 1.9 of bad input wiring, getting progressively worse from (b) to (d), the impedance X-X acts as a source of unwanted input signal due to the other currents flowing in it as well as the input current.

Connection to 0V elsewhere on the pcb

Insufficient control over pc layout is the most usual cause of arrangement (b), especially if auto-routing layout software is used. Most CAD layout software assumes that the 0V rail is a single node and feels itself free to make connections to it at any point along the track. To overcome this, either specify the input return point as a separate node and

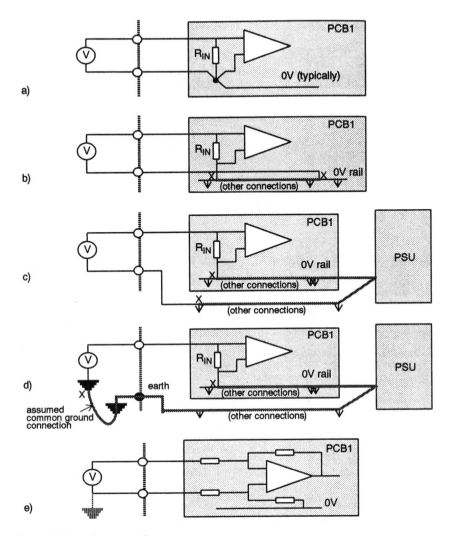

Figure 1.9 Input signal grounding

connect it later, or edit the final layout as required. Manual layout is capable of exactly the same mistake, although in this case it is due to lack of communication between designer and layout draughtsman.

Connection to 0V within the unit

Arrangement (c) is quite often encountered if one pole of the input connector naturally makes contact with the metal case, such as happens with the standard BNC coaxial connector, or if for reasons of connector economy a common ground point is shared between multiple input, output or control signals that are distributed among different boards. With sensitive input signals, the latter is false economy; and if you have to use a BNC-type connector, you can get versions with insulating washers, or mount it on an insulating sub-panel in a hole in the metal enclosure. Incidentally, taking a coax lead internally from an uninsulated BNC socket to the pcb, with the coax outer connected both to the BNC shell and the pcb 0V, will introduce a ground loop (see section 1.1.4) unless it is the only path for ground currents to take.

External ground connection

Despite being the most horrific input grounding scheme imaginable, arrangement (d) is unfortunately not rare. Now, not only are noise signals internal to the unit coupled into the signal path, but also all manner of external ground noise is included. Local earth differences of up to 50V at mains frequency can exist at particularly bad locations such as power stations, and differences of several volts are common. The only conceivable reason to use this layout is if the input signal is already firmly tied to a remote ground outside the unit, and if this is the case it is far better to use a differential amplifier as in Figure 1.9(e), which is often the only workable solution for low-level signals and is in any case only a logical development of the correct approach for single-ended signals (a). If for some reason you are unable to take a ground return connection from the input signal, you will be stuck with ground-injected noise.

All of the schemes of Figure 1.9(b) to (d) will work perfectly happily if the desired input signal is several orders of magnitude greater than the ground-injected interference, and this is frequently the case, which is how they came to be common practice in the first place. If there are good practical reasons for adopting them (for instance, connector or wiring cost restrictions) and you can be sure that interference levels will not be a problem, then do so. But you will need to be very certain that you have control over all possible connection paths before you can be sure that problems won't arise in the field.

1.1.7   Output signal ground

Similar precautions need to be taken with output signals, for the reverse reason. Inputs respond unfavourably to external interference, whereas outputs are the cause of interference. Usually in an electronic circuit there is some form of power amplification involved between input and output, so that an output will operate at a higher current level than an input, and there is therefore the possibility of unwanted feedback.

The classical problem of output-to-input ground coupling is where both input and output share a common impedance, in the same way as the power rail common impedances discussed earlier. In this case the output current is made to circulate through the same conductor as connects the input signal return (Figure 1.10(a)).

A tailor-made feedback mechanism has been inserted into this circuit, by means of R_S. The input voltage at the amplifier terminals is supposed to be V_{in}, but actually it is

$$V_{in}' = V_{in} - (I_{out} \cdot R_S)$$

Redrawing the circuit to reference everything to the amplifier ground terminals (Figure 1.10(b)) shows this more clearly. When we work out the gain of this circuit, it turns out to be

$$V_{out}/V_{in} = A/(1 + [A \cdot R_S/(R_L + R_S)])$$

which describes a circuit that will oscillate if the term $[A \cdot R_S/(R_L + R_S)]$ is more negative than -1. In other words, for an inverting amplifier, the ratio of load impedance to common impedance must be less than the gain, to avoid instability. Even if the circuit remains stable, the extra coupling due to R_S upsets the expected response. Remember also that all the above terms vary with frequency, usually in a complex fashion, so that at high frequencies the response can be unpredictable. Note that although this has been

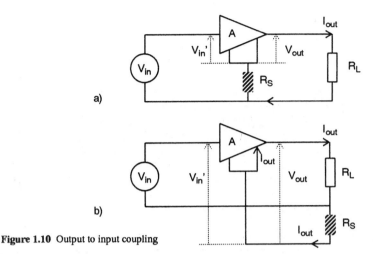

Figure 1.10 Output to input coupling

presented in terms of an analogue system (such as an audio amplifier), any system in which there is input-output gain will be similarly affected. This can apply equally to a digital system with an analogue input and digital outputs which are controlled by it.

Avoiding the common impedance

The preferable solution is to avoid the common impedance altogether by careful layout of input and output grounds. We have already looked at input grounds, and the grounding scheme for outputs is essentially similar: take the output ground return directly to the point from which output current is sourced, with no other connection (or at least, no other susceptible connection) in between. Normally, the output current comes from the power supply so the best solution is to take the return directly back to the supply. Thus the layout of PCB3 in Figure 1.2 should have a separate ground track for the high-current output as in Figure 1.11(a), or the high-current output terminal could be returned directly to the power supply, bypassing PCB3 (b).

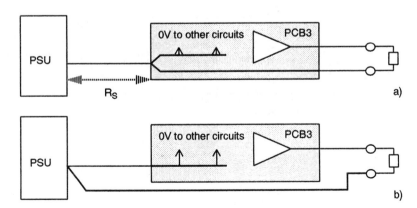

Figure 1.11 Output signal returns

If PCB3 contains only circuits which will not be susceptible to the voltage developed across R_S, then the first solution is acceptable. The important point is to decide in advance where your return currents will flow and ensure that they do not affect the operation of the rest of the circuits. This entails knowing the ac and dc impedance of any common connections, the magnitude and bandwidth of the output currents and the susceptibility of the potentially affected circuits.

1.1.8   Inter-board interface signals

There is one class of signals we have not yet covered, and that is those signals which pass within the unit from one board to another. Typically these are digital control signals or analogue levels which have already been processed, so are not low-level enough to be susceptible to ground noise and are not high-current enough to generate significant quantities of it. To be thorough in your consideration of ground return paths, these signals should not be left out: the question is, what to do about them?

Often the answer is nothing. If no ground return is included specifically for inter-board signals then signal return current must flow around the power supply connections and therefore the interface will suffer all the ground-injected noise V_n that is present along these lines (Figure 1.12). But, if your grounding scheme is well thought out, this may well not be enough to affect the operation of the interface. For instance, 100mV of noise injected in series with a LS-TTL logic interface which has a noise margin of 400mV will have no direct effect. Or, ac noise injection onto a dc analogue signal which is well-filtered at the interface input will be tolerable.

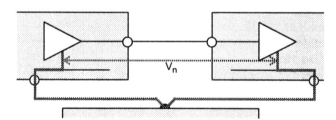

Figure 1.12 Inter-board ground noise

Partitioning the signal return

There will be occasions when taking the long-distance ground return route is not good enough for your interface. Typically these are

- where high-speed digital signals are communicated, and the ground return path has too much inductance, resulting in ringing on the signal transitions;
- when interfacing precision analogue signals which cannot stand the injected noise or low-voltage dc differentials.

If you solve these headaches by taking a local inter-board ground connection for the signal of interest, you run the risk of providing an alternative path for power supply return currents, which nullifies the purpose of the local ground connection. A fraction of the power return current will flow in the local link (Figure 1.13), the proportion depending on the relative impedances, and you will be back where you started.

If you really need the local signal return, but are in trouble with ground return currents, there are two options to pursue:

- separate the ground return (Figure 1.14) for the input side of the interface from the rest of the ground on that pcb. This has the effect of moving the

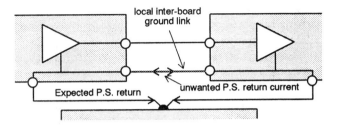

Figure 1.13 Power supply return currents through inter-board links

ground noise injection point inboard, after the input buffer, which may be all that you need. A development of this scheme is to include a "stopper" resistor of a few ohms in the gap X-X. This prevents dc ground current flow because its impedance is high relative to that of the correct ground path, but it effectively ties the input buffer to its parent ground at high frequencies and prevents it from floating if the inter-board link is disconnected.

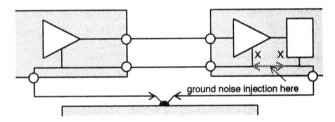

Figure 1.14 Separating the ground returns

• use differential connections at the interface. The signal currents are now balanced and do not require a ground return; any ground noise is injected in common mode and is cancelled out by the input buffer. This technique is common where high-speed or low-level signals have to be communicated some distance, but it is applicable at the inter-board level as well. It is of course more expensive than typical single-ended interfaces since it needs dedicated buffer drivers and receivers.

1.1.9   Star-point grounding

One technique that can be used as a circuit discipline is to choose one point in the circuit and to take all ground returns to this point. This is then known as the "star point", or "circuit mecca" in American terms. Figure 1.2 shows a limited use of this technique in connecting together chassis, mains earth, power supply ground and 0V returns to one point. It can also be used as a local sub-ground point on printed circuit layouts.

When comparatively few connections need to be made this is a useful and elegant trick, especially as it offers a common reference point for circuit measurements. It can be used as a reference for power supply voltage sensing, in conjunction with a similar star point for the output voltage (Figure 1.2 again). It becomes progressively messier as more connections are brought to it, and should not substitute for a thorough analysis of the anticipated ground current return paths.

1.1.10 Ground connections between units

Much of the theory about grounding techniques tends to break down when confronted with the prospect of several interconnected units. This is because the designer often has either no control over the way in which units are installed, or is forced by safety-related or other installation practices to cope with a situation which is hostile to good grounding practice.

The classic situation is where two mains powered units are connected by one (or more) signal cable (Figure 1.15). This is the easiest situation to explain and visualize; actual set-ups may be complicated by having several units to contend with, or different and contradictory ground regimes, or by extra mechanical bonding arrangements.

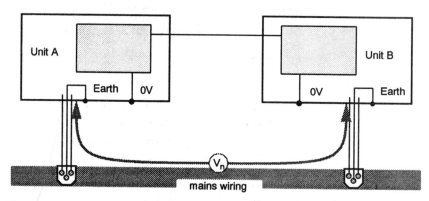

Figure 1.15 Inter-unit ground connection via the mains

This configuration is exactly analogous to that of Figure 1.12. Ground noise, represented by V_n, is coupled through the mains earth conductors and is unpredictable and uncontrollable. If the two units are plugged in to the same mains outlet, it may be very small, though never zero, as some noise is induced simply by the proximity of the live and neutral conductors in the equipment mains cable. But this configuration cannot be prescribed: it will be possible to use outlets some distance apart, or even on different distribution rings, in which case the ground connection path could be lengthy and could include several noise injection sources. Absolute values of injected noise can vary from less than a millivolt rms in very quiet locations to the several volts, or even tens of volts, mentioned in section 1.1.6. This noise effectively appears in series with the signal connection.

In order to tie the signal grounds in each unit together you would normally run a ground return line along with the signal in the same cable, but then

- noise currents can now flow in the signal ground, so it is essential that the impedance of the ground return (R_s) is much less than the noise source impedance (R_n) – usually but not invariably the case – otherwise the ground-injected noise will not be reduced;

- you have created a ground loop (Figure 1.16, and compare this with section 1.1.4) which by its nature is likely to be both large and variable in area, and to intersect various magnetic field sources, so that induced ground currents become a real hazard.

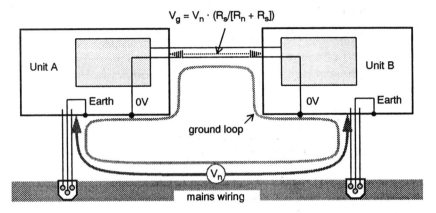

$$V_g = V_n \cdot (R_s/[R_n + R_s])$$

Figure 1.16 Ground loop via signal and mains earths

Breaking the ground link

If the susceptibility of the signal circuit is such that the expected environmental noise could affect it, then you have a number of possible design options.

- float one or other unit (disconnect its mains ground connection), which breaks the ground loop at the mains lead. This is already done for you if it is battery-powered and in fact this is one good reason for using battery-powered instruments. On most mains powered equipment, doing this is not an option because it violates the safety protection.

- transmit your signal information via a differential link, as recommended for inter-board signals earlier. Although a ground return is not necessary for the signal, it is advisable to include one to guard against too large a voltage differential between the units. Noise signals are now injected in common-mode relative to the wanted signal and so will be attenuated by the input circuit's common-mode rejection, up to the operating limit of the circuit, which is usually several volts.

- electrically isolate the interface. This entails breaking the direct electrical connection altogether and transmitting the signal by other means, for instance a transformer, opto-coupler or fibre optic link. This allows the units to communicate in the presence of several hundred volts or more of noise, depending on the voltage rating of the isolation; alternatively it is useful for communicating low-level ac signals in the presence of relatively moderate amounts of noise that cannot be eliminated by other means.

1.1.11   Shielding

Some mention must be made here of the techniques of shielding inter-unit cables, even though this is more properly the subject of chapter 7. Shielded cable is used to protect signal wires from noise pickup, or to prevent power or signal wires from radiating noise. This apparently simple function is not so simple to apply in practice. The characteristics of shielded cable are discussed later (see section 1.2.4); here we shall look at how to apply it.

At which end of a cable do you connect the shield, and to what? There is no one correct answer, because it depends on the application. If the cable is used to connect two units which are both contained within screened enclosures to keep out or keep in RF energy, then the cable shield has to be regarded as an extension of the enclosures and it must be connected to the screening at both ends via a low-inductance connection, preferably the connector screen itself (Figure 1.17). This is a classic application of EMC principles and is discussed more fully in section 8.5 and 8.7. Note that if both of

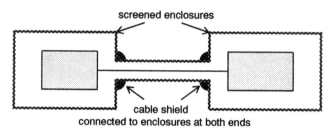

Figure 1.17 RF cable shield connections

the unit enclosures are themselves separately grounded then you have formed a ground loop (again). Because ground loops are a magnetic coupling hazard, and because magnetic coupling diminishes in importance at higher frequencies, this is often not a problem when the purpose of the screen is to reduce hf noise. The difficulty arises if you are screening both against high and low frequencies, because at low frequencies you should ground the shield at one end only, and in these cases you may have to take the expensive option of using double-shielded cable.

The shield should not be used to carry signal return currents unless it is at RF and you are using coaxial cable. Noise currents induced in it will add to the signal, nullifying the effect of the shield. Typically, you will use a shielded pair to carry high-impedance low-level input signals which would be susceptible to capacitive pickup. (A cable shield will *not* be effective against magnetic pickup, for which the best solution is twisted pair.)

Which end to ground

If the input source is floating, then the shield can be grounded at the amplifier input. A source with a floating screen around it can have this screen connected to the cable shield. But, if the source screen is itself grounded, you will create a ground loop with the cable shield, which is undesirable: ground loop current induced in the shield will couple into the signal conductors. One or other of the cable shield ends should be left floating, depending on the relative amount of unavoidable capacitive coupling to ground (C_c) that exists at either end. If you have the choice, usually it is the source end (which may be a transducer or sensor) that has the lower coupling capacitance so this end should be floated.

If the source is single-ended and grounded, then the cable shield should be grounded at the source and either left floating at the (differential) input end or connected through a choke or low value resistor to the amplifier ground. This will preserve dc and low-frequency continuity while blocking the flow of large induced high-frequency currents along the shield. The shield should never be grounded at the opposite end to the signal. Figure 1.18 shows the options.

Electrostatic screening

When you are using shielded cable to prevent electrostatic radiation from output or

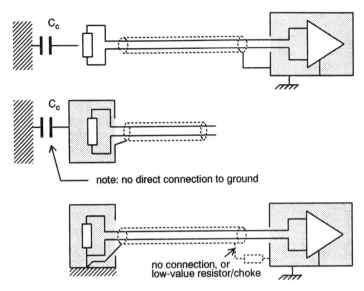

Figure 1.18 Cable shield connection options

inter-unit lines, ground loop induction is usually not a problem because the signals are not susceptible, and the cable shield is best connected to ground at both ends. The important point is that each conductor has a distributed (and measurable) capacitance to the shield, so that currents on the shield will flow as long as there are ac signals propagating within it. These shield currents must be provided with a low-impedance ground return path so that the shield voltages do not become substantial. The same applies in reverse when you consider coupling of noise induced on the shield into the conductors.

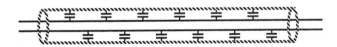

Figure 1.19 Conductor-to-shield coupling capacitance

Surface transfer impedance

At high frequencies, the notion of surface transfer impedance becomes useful as a measure of shielding effectiveness. This is the ratio of voltage developed between the inner and outer conductors of shielded cable due to interference current flowing in the shield, expressed in milliohms per unit length. It should not be confused with characteristic impedance, with which it has no connection. A typical single braid screen will be ten milliohms/m or so below 1MHz, rising at a rate of 20dB/decade with increasing frequency. The common aluminium/mylar foil screens are around 20dB worse. Unhappily, surface transfer impedance is rarely specified by cable manufacturers.

1.1.12 The safety earth

A brief word is in order about the need to ensure a mains earth connection, since it is obvious from the preceding discussion that this requirement is frequently at odds with anti-interference grounding practice. Most countries now have electrical standards which require that equipment powered from dangerous voltages should have a means of protecting the user from the consequences of component failure. The main hazard is deemed to be inadvertent connection of the live mains voltage to parts of the equipment with which the user could come into contact directly, such as a metal case or a ground terminal.

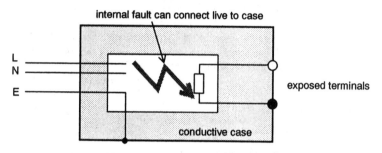

Figure 1.20 The need for a safety earth

Imagine that the fault is such that it makes a short circuit between live and case, as shown in Figure 1.20. These are normally isolated and if no earth connection is made the equipment will continue to function normally – but the user will be threatened with a lethal shock hazard without knowing it. If the safety earth conductor is connected then the protective mains fuse will blow when the fault occurs, preventing the hazard and alerting the user to the fault.

For this reason a safety earth conductor is mandatory for all equipment that is designed to use this type of protection, and does not rely on extra levels of insulation. The conductor must have an adequate cross-section to carry any prospective fault current, and all accessible conductive parts must be electrically bonded to it. The general requirements for earth continuity are

- the earth path should remain intact until the circuit protection has operated;
- its impedance should not significantly or unnecessarily restrict the fault current

As an example, BS415 requires a resistance of less than 0.5Ω at 10A for a minute. Design for safety is covered in greater detail in section 9.1.

1.2 Wiring and cables

This section will look briefly at the major types of wire and cable that can be found within typical electronic equipment. There are so many varieties that it comes as something of a surprise to find that most applications can be satisfied from a small part of the range. First, a couple of definitions: wires are single-circuit conductors, insulated or not; cables are groups of individual conductors, separately insulated and mechanically contained within an overall sheath.

1.2.1 Wire types

The simplest form of wire is tinned copper wire, available in various gauges depending on required current carrying capacity. Component leads are almost invariably tinned copper, but the wire on its own is not used to a great extent in the electronics industry. Its main application is for links on printed circuit boards, but these are going out of fashion as the increasing use of double-sided and multilayer plated-through-hole boards makes them redundant. Tinned copper wire can also be used in re-wirable fuselinks. Insulated copper wire is used principally in wound components such as inductors and transformers. The insulating coating is a polyurethane compound which has self-fluxing properties when heated, which makes for ease of soldered connection, especially to thin wires.

Table 1.1 compares dimensions, current capacity and other properties for various sizes of copper wire. In the UK the wires are specified under BS4109 for tinned copper and BS6811 for insulated, and are sold in metric sizes. BS6811 supersedes BS4520 and many suppliers continue to offer wire to the earlier specification. Two grades of insulation are available, Grade 1 being thinner; Grade 2 has roughly twice the breakdown voltage capability.

Wire inductance

We mentioned earlier that any length of wire has inductance as well as resistance. The approximate formula for the inductance of a straight length of round section wire at high frequencies is

$$L \quad = \quad K \cdot l \cdot (2.3 \log_{10}(4l/d) - 1) \text{ microhenries}$$

where   l and d are length and diameter respectively, l >> d and K is 0.0051 for dimensions in inches or 0.002 for dimensions in cm.

This equation is used to derive the inductance of a 1m length (note that this is not quite the same as inductance per metre) in Table 1.1 and you can see that inductance is only marginally affected by wire diameter. Low values of inductance are not easily obtained by adding cross-section and the reactive component of impedance dominates above a few kiloHertz whatever the size of the conductor. A useful rule of thumb is that the inductance of a one inch length of ordinary equipment wire is around 20nH and that of a one centimetre length is around 7nH. This factor becomes important in high speed digital and rf circuits where performance is limited by physical separation, and also in circuits where the rate-of-change of current (dI/dt) is high.

Equipment wire

Equipment wire, or hook-up wire in American parlance, is classified mainly according to its insulation. This determines the voltage rating and the environmental properties of the wire, particularly its operating temperature range and its resistance to chemical and solvent attack. The standard type of wire, and the most widely available, is PVC insulated to BS4808 which has a maximum temperature rating of 85°C. As well as current ratings at 25°C you will find specifications at 70°C; these allow for a 15°C temperature rise, to the maximum rated temperature, at the specified current. Temperature ratings of 70°C for large conductor switchgear applications and 105°C to American and Canadian UL and CSA standards are also available in PVC. PTFE is used for wider temperature ranges, up to 200°C, but is harder to work with. Other more specialised insulations include extra-flexible PVC for test leads and silicone rubber for high temperature (150°C) and harsh environments. Many wires carry military, telecom

Wire size (mm dia)	1.6	1.25	0.71	0.56	0.315	0.2
Approx. standard wire gauge (SWG)	16	18	22	24	30	35
Approx. American Wire Gauge (AWG)	14	16	21	23	28	32
Current rating (Amps)	22	12.2	3.5	2.5	0.9	0.33
Fusing current (Amps)	70	45	25	17	9	5
Resistance/metre @ 20°C (Ω)	0.0085	0.014	0.043	0.069	0.22	0.54
Inductance of 1 metre length (μH)	1.36	1.41	1.53	1.57	1.69	1.78

Table 1.1 Characteristics of copper wire

Wire size (no. of strands/mm dia)	1/0.6	7/0.2	16/0.2	24/0.2	32/0.2	63/0.2
Resistance (Ω/1000m at 20°C)	64	88	38	25.5	19.1	9.7
Current rating at 70°C (A)	1.8	1.4	3.0	4.5	6.0	11.0
Current rating at 25°C (A)	3.0	2.0	4.0	6.0	10.0	18.0
Voltage drop/metre at 25°C current	192mV	176mV	152mV	153mV	191mV	175mV
Voltage rating	1KV	1KV	1KV	1.5KV	1.5KV	1.5KV
Overall diameter (mm)	1.2	1.2	1.55	2.4	2.6	3.0
Near equivalent American Wire Gauge (not direct equivalent)	23	24	20	18	17	15

Table 1.2 Characteristics of BS4808 equipment wire

	Kynar: 30AWG	26AWG	Tefzel: 30AWG	26AWG
Conductor dia (mm)	0.25	0.4	0.25	0.4
Maximum service temperature °C	105	105	155	155
Resistance/m @ 20°C (W)	0.345	0.136	0.345	0.136
Voltage rating (V)	-	-	375	375
Current rating @ 50°C (A)	-	-	2.6	4.5

Table 1.3 Characteristics of wire-wrap wire

and safety authority approval and have to be specified on projects that are carried out for these customers.

Table 1.2 is included here as a guide to the electrical characteristics of various commonly-available PVC equipment wires. Note that the published current ratings of each wire are related to permitted temperature rise. Copper has a positive temperature coefficient of resistivity of 0.00393 per °C, so that resistance rises with increasing current; the resistance at room temperature may be optimistic by several percent if the actual ambient temperature is high or if significant self-heating occurs.

Cross-sectional area (mm²)	0.5	0.75	1.0	1.25	1.5	2.5
Current-carrying capacity (A)	3	6	10	13	16	25
Voltage drop per amp per metre (mV)	93	62	46	37	32	19
Maximum supportable mass (kg)	2	3	5	5	5	5

Correction factor for ambient temperature

60°C rubber and pvc cables:	Temp.	35°C	40°C	45°C	50°C	55°C
	CF	0.92	0.82	0.71	0.58	0.41
85°C HOFR rubber cables:	Temp.	35-50°C	55°C	60°C	65°C	70°C
	CF	1.0	0.96	0.83	0.67	0.47

Table 1.4 Characteristics of BS6500 mains cables
Source: IEE Wiring Regulations 15th Edition

Wire-wrap wire

A further specialised type of wire is that used for wire-wrap construction. This is available primarily in two sizes, with two types of insulation: Kynar, trademark of Pennwalt, and Tefzel®, trademark of Du Pont. Tefzel is the more expensive but has a higher temperature rating and is easier to strip. Table 1.3 lists the properties of the four types.

1.2.2   Cable types

Ignoring the more specialized types, cables can be divided loosely into three categories:

• power

• data and multicore

• rf

1.2.3   Power cables

Because mains power cables are inherently meant to carry dangerous voltages they are subject to strict standards: in the UK the principal one is BS6500. International ones are IEC227 or IEC245. These standards have been harmonised throughout the CENELEC countries in Europe so that any equipment which uses a cable with a harmonised code number will be acceptable throughout Europe. BS6500 specifies a range of current ratings and allows a variety of sheath materials depending on application. The principal ones are rubber and PVC; rubber is about twice the price of PVC but is somewhat more flexible and therefore suitable for portable equipment, and can be obtained in a high-temperature HOFR (heat and oil resisting, flame retardant) grade. The current-carrying capacities and voltage drops for dc and single-phase ac, and supportable mass are shown in Table 1.4.

Unfortunately, American and Canadian mains cables also need to be approved, but the approvals authorities are different (UL and CSA). Cables manufactured to the European harmonised standards do not meet UL/CSA standards and vice versa. So, if you intend to export your mains-powered equipment both to Europe and North America you will need to supply it with two different cables. The easy way to do this is to use a CEE-22 6 Amp connector on the equipment and supply a different cable set depending on the market. This practice has been adopted by virtually all of the large-volume multinational equipment suppliers with the result that the CEE-22 mains inlet is universally accepted. There are also several suppliers of ready-made cable sets for the different countries!

1.2.4   Data and multicore cables

Multicore cables are used when you need to transport several signals between the same source and destination. They should never be used for mains power because of the hazards that could be created by a cable failure, nor should high-power and signal wires be run within the same cable because of the risks of interference. Conventional multicore is available with various numbers of conductors in 7/0.1mm, 7/0.2mm and 16/0.2mm, with or without an overall braided screen. As well as the usual characteristics of current and voltage ratings, which are less than the ratings for individual wires because the conductors are bunched together, inter-conductor capacitance is an important consideration, especially for calculating crosstalk (to which we return shortly). It is not normally specified for standard multicore, although nominal conductor-to-screen capacitances of 150-200pF/m are sometimes quoted. For a more complete specification you need to use data cable.

Data communication cables

Data cables are really a special case of multicore, but with the explosion in data communications they now deserve a special category of their own. Transmitting digital data presents special problems, notably

- the need to communicate several parallel channels at once, usually over short distances, which has given rise to ribbon cable;

- the need to communicate a few channels of high-speed serial data over long distances with a high data integrity, which has given rise to cables with multiple individually-screened conductor pairs in an overall sheath which may or may not be screened.

Inter-conductor capacitances and characteristic impedances (which we will discuss when we come to transmission lines) are important for digital data transmission and are quoted for most of these types. Table 1.5 summarizes the characteristics of the most common of them.

Shielding and microphony

Shielding of data and multicore falls into three categories:

- copper braid. This offers a good general-purpose electrical shield but cannot give 100% shield coverage - 80-95% is typical - and it increases the size and weight of the cable.

- tape or foil. The most common of these is aluminised mylar. A drain wire is run in contact with the metallisation to provide a terminating contact and to reduce the inductance of the shield when it is helically-wound. This provides a fairly mediocre degree of shielding but hardly affects the size, weight and flexibility of the cable at all.

- composite foil and braid. These provide excellent electrostatic shielding for demanding environments but are more expensive - about twice the price of foil types.

For small-signal applications, particularly low-noise audio work, another cable

Cable type	Ribbon: straight	twisted-pair	Round: Type A	Type B
Inter-conductor capacitance pF/m	50	72	40-115	41-98
Conductor-screen capacitance pF/m	-	-	66-213	72-180
Characteristic impedance Ω	105	105	-	50
Voltage rating V	300	300	300	30

Type A: multi-pair/multicore overall foil screened cable
Type B: multi-pair individually foil screened cable

Table 1.5 Characteristics of data transmission cables

Cable type	URM43	URM67	RG58C/U	RG174A/U	RG178B/U
Overall diameter mm	5	10.3	5	2.6	1.8
Conductor material	Sol 1/0.9	Str 7/0.77	Str 19/0.18	Str 7/0.16	Str 7/0.1
Dielectric material	⊢———	Solid polythene/polyethylene		———⊣	PTFE
Voltage rating *	2.6kV pk	6.5kV pk	3.5kV pk	1.5kV rms	1kV rms
Attenuation dB/10m @100MHz	1.3	0.68	1.6	2.9	4.4
1GHz	4.5	2.5	6.6	10	14
Temperature range °C	⊢———	-40 to +85	———	⊣	-55/+200
Cost per 100m £ †	18.9	70.0	22.5	26.3	81.9

* voltage ratings may be specified differently between manufacturers
† prices are average 1990 costs. Since prices vary with the cost of raw materials these figures should be taken for comparison purposes only

Table 1.6 Characteristics of 50Ω coax cables

property is important – microphony due to triboelectric induction. Any insulator generates a static voltage when it is rubbed against a dissimilar material, and this effect results in a noise voltage between conductor and screen when the cable is moved or vibrated. Special low-noise cable is available which minimises this noise mechanism by including a layer of low resistance dielectric material between braid and insulator to dissipate the static charge. When you are terminating this type of cable, make sure the low resistance layer is stripped back to the braid, otherwise you run the risk of a near short circuit between inner and outer.

1.2.5   RF cables

Cables for the transport of radio frequency signals are almost invariably coaxial, apart from a few specialised applications such as hf aerial feeder which may use balanced lines. Coax's outstanding property is that the field due to the signal propagating along it is confined to the inside of the cable (Figure 1.21), so that interaction with its external environment is kept to a minimum. A further useful property is that the characteristic impedance of coax is easily defined and maintained. This is important for rf applications as in these cases cable lengths frequently exceed the operating wavelength.

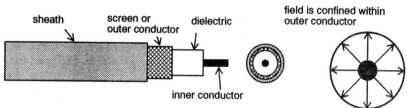

Figure 1.21 Coax cable

The generic properties of transmission lines – of which coax is a particular type – will be discussed in section 1.3. The parameters that you will normally find in coax specifications are as follows:

- characteristic impedance (Z_0): the universal standard is 50Ω, since this results in a good balance between mechanical properties and ease of circuit application. 75Ω and 93Ω are other standards which find application in video and data systems. Any other impedance must be regarded as a special.

- dielectric material. This affects just about every property of the cable, including Z_0, attenuation, voltage handling, physical properties and temperature range. Solid polythene or polyethylene are the standard materials; cellular polyethylene, in which part of the dielectric insulation is provided by air gaps, offers lower weight and lower attenuation losses but is more prone to physical distortion than solid. These two have a temperature rating of 85°C. PTFE is available for higher temperature (200°C) and lower loss applications but is much more expensive.

- conductor material. Copper is universal. Silver plating is sometimes used to enhance high-frequency conductivity through the skin effect, or copper can be plated onto steel strands for strength. Inner conductors can be single or stranded; stranded is preferred when the cable will be subject to flexing. The outer conductor is normally copper braid, again for flexibility. The degree of braid coverage affects high-frequency attenuation and also the shielding effectiveness. Solid outer conductor is available for extreme applications that don't require flexing.

- voltage rating. A thicker cable can be expected to have a higher voltage rating and a lower attenuation. You cannot easily relate the voltage rating to power handling ability unless the cable is matched to its characteristic impedance. If the cable isn't matched, voltage standing waves will exist which will produce peaks at distinct locations along the cable higher than would be expected from the power/impedance relationship.

- attenuation. Losses in the dielectric and conductors result in increasing attenuation with frequency and distance, so attenuation is quoted per 10 metres at discrete frequencies and you can interpolate to find the attenuation at your operating frequency. Cable losses can easily catch you out, especially if you are operating long cables over a wide bandwidth and forget to allow for several extra dB of loss at the top end.

Readily-available coax cables are specified to two standards, the US MIL-C-17 for the RG/U (Radio Government, Universal) series and the UK BS2316 for the UR-M

(Uniradio) series. Table 1.6 gives comparative data for a few common 50Ω types.

One word of warning: never confuse screened audio cable with rf coax. The braids and dielectric materials are quite different, and audio cable's Z_0 is undefined and its attenuation at high frequencies is large. If you try to feed rf down it you won't get much at the other end! On the other hand, rf coax *can* be used to carry audio signals.

1.2.6   Twisted pair

Special mention should be given to twisted pair because it is a particularly effective and simple way of reducing both magnetic and capacitive interference pickup. Twisting the wires tends to ensure a homogeneous distribution of capacitances. Both capacitance to ground and to extraneous sources are balanced. This means that common-mode capacitive coupling is also balanced, allowing high common-mode rejection. Figure 1.22 compares twisted and un-twisted pairs. But note that if your problem is already common-mode capacitive coupling, twisting the wires won't help. For that, you need shielding.

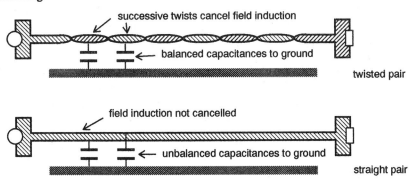

Figure 1.22 The advantage of twisted pair wires

Twisting is most useful in reducing low-frequency magnetic pickup because it reduces the magnetic loop area to almost zero. Each twist reverses the direction of induction so, assuming a uniform external field, two successive twists cancel the wires' interaction with the field. Effective loop pickup is now reduced to the small areas at each end of the pair, plus some residual interaction due to non-uniformity of the field and irregularity in the twisting. Assuming that the termination area is included in the field, the number of twists per unit length is unimportant: around 8–16 turns per foot (26 – 50 turns per metre) is usual. Figure 1.23 shows measured magnetic field attenuation versus frequency for twisted 22AWG wires compared to parallel 22AWG wires spaced at 0.032".

A further advantage of twisting pairs together is that it allows a fairly reproducible characteristic impedance. When combined with an overall shield to reduce common-mode capacitive pickup, the resulting cable is very suitable for high-speed data communication as it reduces both radiated noise and induced interference to a minimum.

1.2.7   Crosstalk

When more than one signal is run within the same cable bundle for any distance, the mutual coupling between the wires allows a portion of one signal to be fed into another,

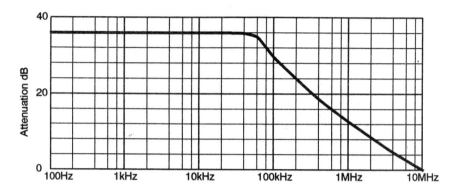

Figure 1.23 Magnetic field attenuation of twisted pair   *Source:*
"Unscrambling the mysteries about twisted wire", R.B.Cowdell, IEEE EMC Symposium 1979, p.183

and vice versa. The phenomenon is known as crosstalk. Strictly speaking, crosstalk is not only a cable phenomenon but refers to any unwanted interaction between nominally un-coupled channels. The coupling can be predominantly either capacitive, inductive, or due to transmission-line phenomena.

The equivalent circuit for capacitive coupling at low-to-medium frequencies where the cable can be considered as a lumped component (in contrast to high frequencies where it must be considered as a transmission line) is as shown in Figure 1.24.

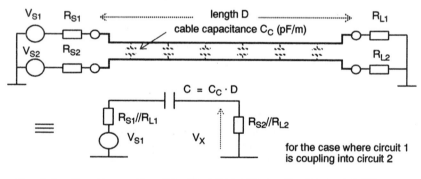

Crosstalk voltage $V_X = V_{S1} \cdot \{(R_{S2}//R_{L2}) / [(R_{S2}//R_{L2}) + (R_{S1}//R_{L1}) + (1/\omega C)]\}$

Figure 1.24 Crosstalk equivalent circuit

In the worst case where the capacitive coupling impedance is much lower than the circuit impedance, the crosstalk voltage is determined only by the ratio of circuit impedances.

Digital crosstalk

Crosstalk is well-known in the telecomms and audio worlds, for example where separate speech channels are transmitted together and one breaks through onto another, or where stereo channel separation at high frequencies is compromised. Although digital data might seem at first sight immune from crosstalk, in fact it is a serious threat to data integrity as well. The capacitive coupling is all but transparent to fast edges with

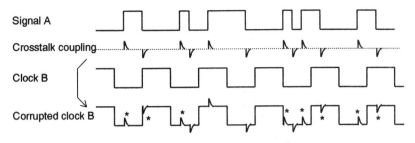

Figure 1.25 Digital crosstalk effects

the result that clocked data can be especially corrupted, as Figure 1.25 shows. If the logic noise immunity is poor, severe mis-clocking can result. A couple of worked examples will demonstrate the nature of the problem.

(a) Two audio circuits with 10kΩ source and load impedances are run in 2 metres of multicore cable with interconductor capacitances of 150pF/m. What is the crosstalk ratio at 10kHz?

The coupling capacitance C_C is 2 metres of 150pF/m = 300pF. At 10kHz this has an impedance of 53kΩ.

The source and load impedances in the crosstalk circuit in each case are 10K//10K = 5kΩ.

So the crosstalk will be

5K / (5K + 5K + 53K)   =   **22dB**: unacceptable in just about any situation!

If the output drive impedance is reduced from 10kΩ to 50Ω then the crosstalk becomes

49 / (49 + 49 + 53K)   =   **60dB**

which is acceptable for many purposes, though probably not for hi-fi.

(b) Two EIA-232D (RS-232) serial data lines are run in 16m of data cable (not individual twisted pair) which has a core/core capacitance of 108pF/m. The transmitters and receivers conform to the EIA-232 spec of 300Ω output impedance, 5kΩ input impedance, ±10V swing and 30V/μsec rise time. What is the expected magnitude of interference spikes on one circuit due to the other?

Coupling capacitance here is 16 x 108pF   =   1728pF.

The current that will be flowing after t seconds in an RC circuit fed from a ramping voltage with a constant dV/dt is

$$I \quad = \quad C \, dV/dt \, (1 - \exp[-t/RC])$$

which for our case with dV/dt = 30V/μsec for 0.66 μsec and a circuit resistance of 567Ω is 25mA. This translates to a peak voltage across the load resistance of (300//5K//5K) of

$$25. \, 10^{-3} \times 267 \quad = \quad \textbf{6.8V}$$

This is one reason why EIA-232 isn't suitable for long distances and high data rates!

Crosstalk can be combated with a number of strategies, which follow from the above examples. These are

- reduce the circuit source and/or load impedances. Ideally, the offending circuit's source impedance should be high and the victim's should be low. Low impedances require more capacitance for a given amount of coupling.

- reduce the mutual coupling capacitance. Use a shorter cable, or select a cable with lower core-to-core capacitance per unit length. Note that for fast or high-frequency signals this won't solve anything, because the impedance of the coupling capacitance is lower than the circuit impedances. If you use ribbon cable, sacrifice some space and tie a conductor to ground between each signal conductor; another alternative is ribbon cable with an integral ground plane. Best of all, use an individual screen for each circuit. The screen must be grounded or you gain nothing at all from this tactic!

- reduce the signal circuit bandwidth to the minimum required for the data rate or frequency response of the system. As can be seen from (b) above, the coupling depends directly on the rise time of the offending signal. Slower risetimes mean less crosstalk. If you do this by adding a capacitance in parallel with the input load resistor (across R_{L2} in Figure 1.24) this will act as a potential divider with the core-to-core capacitance, as well as reducing the input impedance for high-frequency noise.

- use differential transmission. The bogey of crosstalk is a major reason for the popularity of differential data standards such as EIA-422 (RS-422) at high data rates. Coupling capacitance is not necessarily reduced by using paired lines, but the crosstalk is now injected in common-mode and so benefits from the common-mode rejection of the input buffer. The limiting factor to the degree of rejection that can be obtained is the unbalance in coupling capacitance of each half of the pair. This is why twisted pair cable is advised for differential data transmission.

1.3 Transmission lines

Electronics is not a homogeneous discipline. It tends to divide into set areas: analogue, digital, power, rf and microwave. This is a pragmatic division because different mathematical tools are used for these different areas and it is rare for any one designer to be proficient in all or even most of them. Unhappily for the designer, nature knows nothing of these civilised distinctions; all electrons follow the same physical laws regardless of who observes them and regardless of their speed.

When signal frequencies are low, it is possible to imagine that circuit operation is constrained by the laws of circuit theory: Thévenin, Kirchoff *et al.* This is not actually true. Electrons do not read circuit diagrams, and they operate according to the rather grander and more universal laws of Electromagnetic Field Theory, but the difference at low frequencies is so slight that circuit-theoretical predictions are indistinguishable from the real thing. Circuit theory serves electronic engineers well.

As the speed of circuit operation rises, though, it breaks down. It is not that electrons change their behaviour at higher frequencies; there is no cut-off point beyond which everything is different. It is simply that the predictions of circuit theory diverge from those of Electromagnetic Field theory, and the latter, having the backing of nature, wins. One of the consequences of this victory is that perfectly ordinary lengths of wire and cable magically turn into transmission lines.

Transmission line effects

There is no straightforward answer to the question "when do I have to start considering transmission line properties?" The best response is, when the effects become important to you. One of the simplest electrical laws is that which relates frequency, wavelength and the speed of light:

$$\lambda \quad = \quad 3 \cdot 10^8 / f$$

which is modified because of the reduction in velocity of propagation when a dielectric medium is involved by the relative permittivity or dielectric constant of the medium,

$$\lambda_d \quad = \quad \lambda / \sqrt{\varepsilon_r}$$

One rule of thumb is that a cable should be considered as a transmission line when the wavelength of the highest frequency carried is less than ten times its length. You may be embarrassed by transmission line effects at lengths of one fortieth the wavelength or less if you are working with precision high-speed signals, or you may not care until the length reaches a quarter wavelength – though by then you will certainly be getting some odd results.

Critical lengths for pulses

If as a digital engineer you work in terms of risetimes rather than frequency, then a roughly equivalent rule of thumb is that if the shortest risetime is less than three times the travelling time along the length of the cable you should be thinking in terms of transmission lines. Thus for a rise time of 10ns in coax with a velocity factor $(1/\sqrt{\varepsilon_r})$ of 0.66 the critical length will be two thirds of a metre.

1.3.1   Characteristic impedance

Characteristic impedance (Z_0) is the most important parameter for any transmission line. It is a function of geometry as well as materials and it is a dynamic value independent of line length; you can't measure it with a multimeter. It is related to the conventional distributed circuit parameters of the cable or conductors by

$$Z_0 \quad = \quad [(R + j\omega L) / (G + j\omega C)]^{0.5}$$

where   R is the series resistance per unit length (Ω/m)
         L is the series inductance (H/m)
         G is the shunt conductance (mho/m)
         C is the shunt capacitance (F/m)

L and C are related to the velocity factor by

velocity of propagation $= \quad 1/\sqrt{LC} \quad = \quad 3 \cdot 10^8 / \sqrt{\varepsilon_r}$

For an ideal, lossless line the R and G terms are zero and Z_0 reduces to $\sqrt{(L/C)}$. Practical lines have some losses which attenuate the signal, and these are quantified as an attenuation factor for a specified length and frequency (Table 1.6 shows these for coaxial cables). Table 1.7 summarizes the approximate characteristic impedances for various geometries, along with velocity factors of some common dielectric materials. The value 377 (120π) crops up several times: it is a significant number in electromagnetism, being the *impedance of free space* (in ohms), which relates electric and magnetic fields in free-field conditions.

Driving a signal down a transmission line provides an important exception to the

1. Side-by-side parallel strip

$$Z_o = 120/\sqrt{\varepsilon_r} \cdot \ln\{h/w + [(h/w)^2 - 1]^{0.5}\}$$

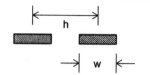

2. Face-to-face parallel strip

$$Z_o = 377/\sqrt{\varepsilon_r} \cdot h/w \quad \text{if } h > 3t, w \gg h$$
$$120/\sqrt{\varepsilon_r} \cdot \ln 4h/w \quad \text{if } h \gg w$$

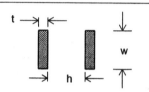

3. Parallel wire

$$Z_o = 120/\sqrt{\varepsilon_r} \cdot \ln\{h/d + [(h/d)^2 - 1]^{0.5}\}$$
$$120/\sqrt{\varepsilon_r} \cdot \ln 2h/d \quad \text{if } d \ll h$$

(Z_o of typical pvc-insulated pairs and twisted pairs is around 100Ω)

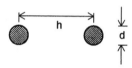

4. Wire parallel to infinite plate

$$Z_o = 60/\sqrt{\varepsilon_r} \cdot \ln\{2h/d + [(2h/d)^2 - 1]^{0.5}\}$$
$$60/\sqrt{\varepsilon_r} \cdot \ln 4h/d \quad \text{if } d \ll h$$

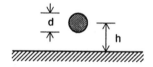

5. Strip parallel to infinite plate

$$Z_o = 377/\sqrt{\varepsilon_r} \cdot h/w \quad \text{if } w > 3h$$
$$60/\sqrt{\varepsilon_r} \cdot \ln 8h/w \quad \text{if } h > 3w$$

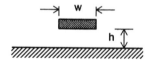

6. Coaxial

$$Z_o = 60/\sqrt{\varepsilon_r} \cdot \ln (D/d)$$

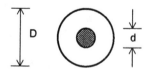

Dielectric constants of various materials	ε_r	Velocity factor ($1/\sqrt{\varepsilon_r}$)
Air	1.0	1.0
Polythene/Polyethylene	2.3	0.66
PTFE	2.1	0.69
Silicone Rubber	3.1	0.57
PVC	5.0	0.45
FR4 Fibreglass PCB	5.0 (typ)	0.45

Table 1.7 Characteristic impedance, geometry and dielectric constants

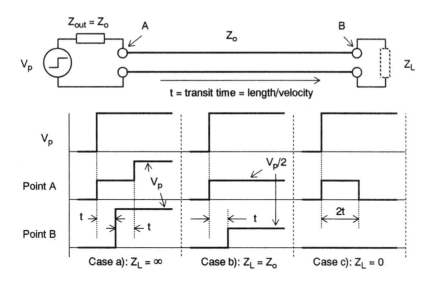

Figure 1.26 Voltage edge propagating along a transmission line

general rule of circuit theory (for voltage drives) that the driving source impedance should be low while the receiving load impedance should be high. When sent down a transmission line, the signal is only received undistorted if both source and load impedances are the same as the line's characteristic impedance. This is said to be the *matched* condition. It is easiest to consider the effects of matching and mis-matching in two parts: in the time domain for digital applications and in the frequency domain for analogue radio frequency applications.

1.3.2   Time domain

Imagine a step waveform being launched into a transmission line from a generator which is matched to the line's characteristic impedance Z_0. We can view the waveform at each end of the line and, because of the finite velocity of propagation down the line, the two waveforms will be different. The results for three different cases of open, matched and short line terminating impedance (these are the easily-visualized special cases) are shown in Figure 1.26. If you have a reasonably fast pulse generator, a wide bandwidth oscilloscope and a length of coax cable you can perform this experiment on the bench yourself in five minutes.

A matched transmission line is actually a simple form of delay line, with delays of the order of tens of nanoseconds achievable from practical lengths. Discrete-component delay lines are smaller but work on the same principle, with the distibuted L and C values being replaced by actual components.

In all cases the long-term result is as would be expected from conventional circuit theory: an open circuit results in V_p, a short circuit results in zero and anything in between results in the output being divided by the potential divider $Z_L/(Z_{out} + Z_L)$, giving $V_p/2$ for the matched case. While the edge is in transit the driving waveform is different.

Forward and reflected waves

Transmission line theory explains the results in terms of a forward and a reflected wave, the two components summing at each end to satisfy the boundary conditions: zero

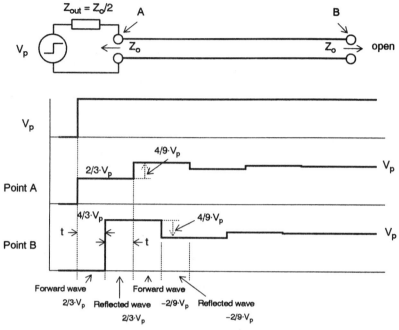

Figure 1.27 A mismatched transmission line

current for an open circuit, zero voltage for a short. Thus in the short-circuit case, the forward wave of amplitude $V_p/2$ generates a reflected wave of amplitude $-V_p/2$ when it reaches the short, which returns to the driving end and sums with the already-existing $V_p/2$ to give zero. In the general case, the ratio of reflected to forward wave amplitude is

$$V_r/V_i \quad = \quad [Z - Z_o] / [Z + Z_o]$$

This explanation is most useful when you want to consider mismatches at both ends. Forward and reflected waves are then continually bounced off each mis-matched end. Take as another example a drive impedance of $Z_o/2$ and an open-circuit load, which is a very crude approximation to an HCMOS logic buffer driving an unterminated HCMOS input. This is shown in Figure 1.27.

Ringing

The reflected wave from the open circuit end now gets reflected in turn from the mis-matched driver end with a lower amplitude, which is reflected back by the open circuit which gets reflected again from the driver with a lower amplitude... Eventually the reflections die away and equilibrium is reached. The waveforms at both ends show considerable "ringing". If you work with digital circuits you will be familiar with ringing if you have ever observed your signals over a few inches of pc track with a fast oscilloscope. The amplitude of the ringing depends entirely on the degree of mismatch between the various impedances, which are complex and for practical purposes essentially unknowable, and the period of the ringing depends on the transit time from driver to termination and hence on line length. A typical ringing frequency for a 0.6mm track over a ground plane on 1.6mm epoxy-glass pcb is 35MHz divided by the line length in metres.

The Bergeron diagram

An accurate determination of the amplitude of the reflections at both ends of a transmission line can be made using a Bergeron diagram. This shows the characteristic impedance of a transmission line as a series of load lines on the input and output characteristics of the line driver and receiver. Each load line originates from the point at which the previous load line intersects the appropriate input/output characteristic. To properly use the Bergeron diagram, you need to know the device characteristics both within and outside the supply rail voltage levels, since ringing carries the signal line voltage outside these points. Many manufacturers of high-speed logic ICs detail its use in their application notes.

Ringing in digital circuits is always undesirable since it leads to spurious switching, but it can be tolerated if the amplitudes involved are within the logic family's noise immunity band, or if the transit times are faster than its response speed. In fact the idealised example in Figure 1.27 shows a step edge which is unrealistic, as practical rise times will damp the response. The only way to avoid it completely is to consider every interconnection as a transmission line, and to terminate each end with its correct characteristic impedance. Very fast circuits are designed in exactly this way; most of us will only meet the problem in severe form when driving long cables.

The uses of mismatching

Mismatching is not always bad. For instance, a very fast, stable pulse generator can be built by feeding a fast risetime edge into a length of transmission line shorted at the far end (Figure 1.28), and taking the output from the input to the line. A 1m length of coax with velocity factor 0.66 will give a 10ns pulse.

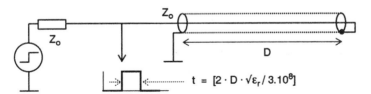

Figure 1.28 Pulse generation with a shorted transmission line

1.3.3   Frequency domain

If you are more interested in radio frequency signals than in digital edges you want to know what a transmission line does in the frequency domain. Consider the transmission line of Figure 1.26 being fed from a continuous sine-wave generator of frequency f and matched to the line's Z_0. Again, the energy can be thought of as a wave propagating along the line until it reaches the load; if the load impedance is matched to Z_0 then there is no reflection and all the power is transferred to the load.

If the load is mismatched then a portion of the incident power is reflected back down the line, exactly like an applied pulse edge. A short or open circuit reflects all the power back. But the signal that is reflected is a continuous wave, not a pulse; so the voltage and current at any point along the line is the vector sum of the voltages and currents of the forward and reflected waves, and depends on their relative amplitudes and phases. The voltage and current distribution down the length of the line forms a so-called "standing wave". The standing wave patterns for four conditions of line termination are shown in Figure 1.29. You can verify this experimentally with a length of fairly leaky coax and a "sniffer" probe, connected to a rf voltmeter, held close to and moved along the coax.

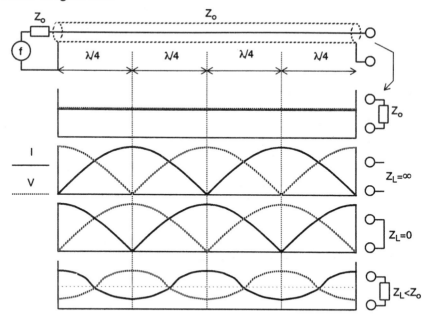

Figure 1.29 Standing waves along a transmission line

Standing wave distribution vs. frequency

Note that the standing wave distribution depends on the wavelength of the applied signal and hence on its frequency. Standing waves at one frequency along a given length of line will differ from those at another. The standing wave pattern repeats itself at multiples of $\lambda/2$ along the line. The amplitude of the standing wave depends on the degree of mismatch, which is represented by the reflection coefficient r, the ratio of reflected current or voltage to incident current or voltage. Standing wave ratio (s.w.r.) is the ratio of maximum to minimum values of the standing wave and is given by

$$\text{s.w.r.} \quad = \quad (1+r)/(1-r) \quad = \quad R_L/Z_o \text{ for a purely resistive termination}$$

Thus a s.w.r. of 1:1 describes a perfectly matched line; infinite s.w.r. describes a line terminated in a short or open circuit. The generator source impedance has no effect on the s.w.r. It depends only on the nature of the load at the far end.

Impedance transformation

The variation of voltage and current along a mismatched transmission line is of great interest to the designers and operators of radio transmitters because it affects the efficiency of power transfer from the transmitter, through the feeder line to the antenna. It is also useful to high-frequency circuit designers as a means of making impedance transformations. Remembering that impedance is voltage divided by current, you can see from Figure 1.29 that the impedance at any given point along the line varies considerably depending on the distance from the termination. For each quarter wavelength, the impedance varies from minimum to maximum. In fact, for a quarter-wave transmission line transformer, the impedance transformation is given by

$$Z_{in} = Z_o^2 / Z_L$$

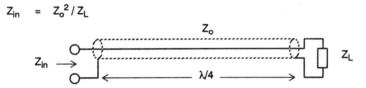

Figure 1.30 Quarter-wave transformer

This useful property is of course frequency dependent; it only occurs at $\lambda/4$ and multiples thereof. If the frequency is changed then the line length departs from $\lambda/4$ and Z_{in} becomes reactive. A related property is that at even multiples of $\lambda/4$ (equivalent to saying any multiple of $\lambda/2$) the original load impedance is regained, whatever the value of Z_o. Thus a shorted line will have a virtually zero impedance $\lambda/2$ away from the short, which property can be used to create a distributed tuned circuit.

Lossy lines

The preceding discussion assumes zero-loss lines, which in practice are unrealisable. For short line lengths the losses are usually insignificant – see Table 1.6 for typical coax losses. Note that these are quoted for the matched condition. If the line is operated with standing waves, the loss is greater than if it is matched because increased voltages and currents are present and the average heat loss is greater for the same power output. The effect of attenuation in long lines is to cause an improvement in s.w.r. towards the generator, since the effect of a mismatch is attenuated in both directions. In the limit, a long cable can make a very good power attenuator!

Chapter 2

Printed Circuits

Virtually every electronic circuit built nowadays uses a printed circuit board (pcb) as both its interconnecting medium and mechanical mounting substrate. The pcb is custom-designed for the circuit it carries and its selection is an important part of the circuit designer's task, though this is frequently delegated to the (perceived) inferior function of pc layout draughtsman. The design of the pcb has a strong effect on the mechanical and electrical performance of the final product. This chapter looks at the main factors that you should consider when working with your new pcb design.

2.1   Board types

An unprocessed board is a laminate of a conductive material and an insulating dielectric substrate. The different materials that are used, and the different ways of laminating and interconnecting between conductive layers, decide the type.

2.1.1   Materials

The conductive layer is almost invariably copper foil, adhesive bonded under heat and pressure to the substrate. The copper cladding thickness is usually specified by its weight per square foot, the most common being one- or two-ounce, other thicknesses being 0.25, 0.5, 3 and 4-oz. Thickness of 1-oz copper is typically 0.035mm ±0.002mm, other weights being thicker or thinner *pro rata*. The main deciding factor in choosing copper weight is its resistivity. Figure 2.1 gives resistance versus track width for the different weights.

The most common laminates are phenolic paper and epoxy glass. Phenolic paper (or synthetic resin bonded paper, s.r.b.p.) is cheaper and can be punched readily, so its major application is in high-volume domestic and other non-critical sectors. It is electrically inferior to epoxy glass, is mechanically brittle, has a poor temperature range, absorbs moisture readily and is unsuitable for plated-through-hole construction.

Epoxy-glass

Epoxy resin with woven glass cloth reinforcement is used for most plated-through-hole (PTH) and multilayer boards. It can also be used for simpler constructions if its better mechanical and electrical properties are needed. It offers much better dimensional stability and is stronger than phenolic paper but this does mean that high-volume board cost-effectiveness is less, because it must be drilled rather than punched. It can be used at temperatures of up to 115°C. The most widely used thickness is 1.6mm, but thicknesses from 0.1mm upwards can be obtained. The common designation of "FR4" refers to the American NEMA specification for flame-retardant epoxy-woven glass board, which is offered by most laminate manufacturers. Flame retardant grades are available for all common materials, and are shown by the manufacturers' ID being marked in red.

For flexible boards the base films are polyester or polyimide. Polyester is cheap but cannot easily be soldered because of its low softening temperature, so is used primarily for flexible "tails". Polyimide is more expensive but components can be mounted on it.

2.1.2   Type of construction

Most circuit requirements can be accommodated on one of the following board types, which are listed roughly by increasing cost.

1. Single-sided

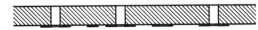

Cheap and cheerful, for simple, low-performance and/or high-volume circuits

2. Double-sided

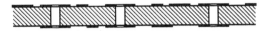

As above but different track pattern on each side of the board, through connections made on assembly; e.g. low density digital applications

3. Flexible

Base material thin and flexible, may be covered with further layers to protect track pattern. May be through-hole plated. Replacement for wiring harnesses

4. Double-sided, plated-through-hole (PTH)

Similar to ordinary double sided but hole barrels are metal plated to interconnect both sides, so different production technique. Medium-to-high density general industrial etc. applications

5. Rigidised Flexible

As flexible, but part-stiffened by rigid plate for component mounting. Used where application requires few components with flexibility

6. Multi-layer

Several layers of base material laminated into single unit. Variants may
have holes passing through all layers or only internal layers (buried
vias). Internal two layers may be power and ground planes. Expensive,
but very high densities achievable

7. Flexi-rigid

Multilayer board with some rigid layers replaced by flexible (usually
polyimide) which extend away from the rigid section to form tails or
hinges. Several rigid areas may be interconnected by flexible to allow
folded shapes for dense packing.

Many variants of the basic multi-layer and flexi-rigid principles are possible, but
their expense usually rules them out for commercial applications unless the overall
assembled board value is very high or unless the application can't be met any other way.
Other techniques and materials are available for specialised functions such as switch
contacts, motor assemblies or microwave systems. If you have these kinds of
applications you will be talking to pcb manufacturers at the concept stage. But, if your
application is more typical, how do you select the best approach?

2.1.3   Choice of type

As in any design problem, many factors need to be balanced to achieve the best final
result. The most important are cost, packing density and electrical performance, with
other factors appearing to a lesser extent.

- *cost.* The above list gives an idea of ranking in terms of bare-board cost,
  though there will be a degree of overlap. The actual cost formula involves
  board quantity ordered, number of processes, number and variety of drill
  holes and the raw material cost as its main parameters. But as well as bare-
  board cost you should also consider the possible effect on overall unit costs:
  the choice of board type can affect assembly, test, repair and rework. For
  instance, if you expect to rework a significant number of units this would
  militate against phenolic paper because of its poorer copper-laminate
  adhesion. At the other extreme, it could rule out multilayer boards because
  of the danger of destroying an otherwise good board by damaging one
  through hole.

- *space limitations.* If your board size is fixed and so is your circuit package
  count, then you have automatically determined packing density and to a
  great extent the optimum board type for lowest cost. Non-PTH boards give
  the lowest packing density. Double-sided PTH can offer between 4 and 7
  cm^2 per 16-pin dual-in-line package depending on track spacing and
  dimensions, while multilayers can approach the practical limit of 2 cm^2 per
  pack. Multilayers are also the only way to fully realise the space advantages

of flat-pack or surface mount components. Discrete leaded components (resistors, capacitors, transistors etc.) reduce the advantage of multilayers because they effectively offer more area for tracking. If there is no limitation on board size, then using a larger, cheaper, low-density board must be balanced against the cost of its larger housing.

- *electrical characteristics.* Phenolic paper may not have sufficiently high bulk resistance, voltage breakdown or low dielectric loss; or the laminate thickness may be determined by required track characteristics, although usually it's easier to work the other way round; or thicker copper may be required for low resistivity. If surface conductivity is likely to be a problem, conformal coating is one solution (see section 2.3.2). If good screening or power/ground plane distribution is required, a four-layer multilayer construction may be a preferable solution, or for lower densities it may be necessary to move up from single-sided to double-sided.

- *mechanical characteristics.* Weight, stiffness and strength may all be important. If you need good resistance to vibration or bowing, either a thicker laminate or stiffening bars may be required. Usually, strength is not critical unless the board carries very heavy components such as a large transformer, in which case epoxy-glass is essential. Coefficient of thermal expansion and maximum temperature rating may need checking if you have a wide-temperature range application.

- *availability.* Just about any pcb manufacturer should be able to cope with ordinary single sided and double sided boards, though prices can vary widely. Most can now offer PTH types. As you progress towards the more exotic flexible and multilayer constructions your options will diminish and you could become locked in to a single source or face unacceptable delivery times. Also, designing complex multilayer boards requires an advanced level of skill, even with improving CAD systems, and your design resources may not cover this when faced with short timescales.

- *reliability and maintainability.* These factors normally favour uncomplicated construction with high-quality materials, for which double-sided PTH on epoxy-glass wins out easily.

- *rigid versus flexi.* The flexi-rigid construction may offer lower assembly costs and better packing density and reliability by eliminating inter-board wiring and connectors. Against this it is more expensive and more prone to difficulties in supply, and may run counter to a repair philosophy based on modular replacement.

2.1.4   Choice of size

If you have a free hand in selecting the size of the board or boards in your system then another set of factors need to be considered. It may be best to go for a standard size such as Eurocard (100 x 160mm) or double-Eurocard (233.4 x 160mm) especially if the boards will be fitted into a modular rack system. Modularizing the total system like this may not be optimum, though; for smaller systems a single large board will do away with the cost of interconnecting components, and will be cheaper to produce than several smaller boards of equivalent area. On the other hand, board material costs depend to some extent on wastage in cutting from the stock laminate, and larger boards are likely to waste more unless their dimensions are matched to the stock. Also, there is a practical limit to the size of very large boards, fixed by considerations of stiffness, dimensional tolerancing and handling, not to mention the capability of the board manufacturer, photoplotter and layout generation. Your own assembly department will also have limits on the board size they can handle. Optimum size for large boards is generally around 30 - 50cm on the longer edge.

Sub-division boundaries

If you are going to split the system into several smaller boards, then position the splits to minimise the number of interconnections. Often this corresponds with splitting it into functional sub-assemblies, and this helps because it is easiest to test each assembled board as a fully functioning sub-unit. Test cost per overall system will decrease as the board size increases, but the expense of the required automatic test equipment will rise. Allow a safety factor in your calculations of the board area needed for each sub-system: subsequent circuit modifications nearly always increase component count, and very rarely reduce it.

2.2 Design rules

Most firms that use pcbs have evolved a set of design rules for layout that have two aims:

- ease of manufacture of the bare board, which translates into lower cost and better reliability;
- ease of assembly, test, inspection and repair of the finished unit, which translates as above.

These rules are necessary to ensure that layout designers know the limits within which they can work, and so that some uniformity of purchasing policy is possible. It has the advantage that the production and servicing departments are faced with a reasonably consistent series of designs emanating from the design department, so that investment in production equipment and training is efficiently used. At the same time, the rules should be reviewed regularly to make sure that they don't unnecessarily restrict design freedom in the light of board production technology advances. Any company, for instance, whose design rules still specify minimum track widths of 0.5mm is not going to achieve the best available packing densities. The design rules should not be enforced so rigorously that they actually prevent the optimum design of a new product.

BS6221 Part 3, "Guide for the design and use of printed wiring boards", presents a good overview of recommended design practices in pc layout and can be used to form the basis of in-house design rules.

Factors in board design which should be considered in the design rules are

- track width and spacing
- hole and pad diameter
- track routing
- ground distribution
- package placement
- component identification
- solder mask

- terminations and connections

2.2.1   Track width and spacing

The minimum width and spacing determine to a large extent the achievable packing density of a board layout. Minimum width for tracks that do not carry large currents is set by the controllability of the etching process, which may vary between board manufacturers and will also, at the margins, affect the cost of the board. You will need to check with your board supplier what minimum width he is capable of processing. 0.3mm should be within most capabilities; 0.15mm is possible but borderline. The narrower widths may be acceptable in isolated instances, such as between IC pads, but are harder to maintain over long distances. Figure 2.3 shows the trade-off in track width and spacing versus number of tracks that can be squeezed between IC pads.

Electrical considerations may affect allowable track width in particular circumstances. Table 1.7 shows the characteristic impedance for various conductor geometries; for a copper track over a ground plane on a double-sided or multilayer board these work out to around 100–150Ω for narrow tracks on 1.6mm thickness, reducing to around 75–100Ω on 0.7mm. If your design includes logic with fast rise-times and you have to interconnect the ICs over appreciable distances (the velocity factor of 0.45 for FR4 board means that an edge will only travel 13cm in a nanosecond) then the tracks must be treated as transmission lines.

Conductor resistance

Track width also, rather more obviously, affects resistance and hence voltage drop for a given current. Figure 2.1 shows the theoretical resistance for various thicknesses of copper track per centimetre. It is derived from the equation

$$R \quad = \quad \rho l / A$$

where ρ is the conductor resistivity

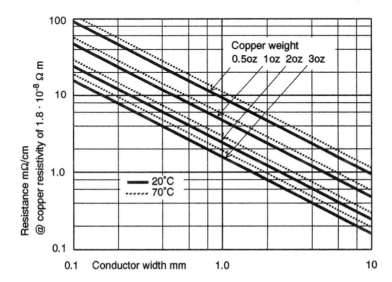

Figure 2.1 Resistance of copper tracks

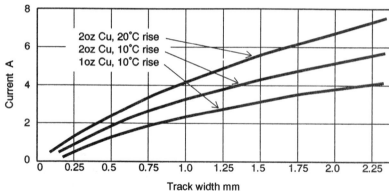

Figure 2.2 Safe currents in pcb tracks
Source: Abstracted from BS6221 : Part 3 : 1984

I is the length
A is the track cross-sectional area

These figures should only be taken as a guide because the actual manufactured tolerances, including base copper, plating and tin-lead thickness will be very wide, amounting to a two-to-one variation in final value. The temperature coefficient of copper means that resistance will vary by several percent over typical ambient temperature ranges, and with self-heating. Plated-through-holes of greater than 0.8mm diameter present less than a milliohm resistance.

Maximum current-carrying ability is determined by self-heating of the tracks. Figure 2.2 shows the safe current versus track width for a given temperature rise.

Voltage breakdown and crosstalk

Track spacing is also determined by production capability and electrical considerations. Minimum spacings similar to track widths (0.15 - 0.3mm) are achievable by most pcb manufacturers. Crosstalk and voltage breakdown are the electrical characteristics which affect spacing. For a benign environment - dry and free from conductive particles - a spacing of 1mm per 200V, allowing for manufacturing tolerances, is adequate. BS6221 part 3 gives greater detail. When mains voltages are present, spacing is normally set by safety approval requirements. Spacings less than 0.5mm risk solder bridging during wave soldering, depending on the transport direction, if solder resist is not used.

Crosstalk (see section 1.2.7) is likely to be the limiting factor on low-voltage digital or high-speed analogue boards. The mechanism is similar to that for cables, but calculating track-to-track capacitance is difficult, and once the track is considered as a transmission line does not yield meaningful results anyway. The simplest rule of thumb is that a track spacing greater than 1mm will result in crosstalk voltages less than 10% of signal voltages for most board configurations. Electrically short connections can be spaced much closer than this without undue concern. Crosstalk can be reduced by routing ground conductors between pairs of signal lines considered susceptible.

2.2.2   Hole and pad diameter

The diameter for component mounting holes should be reasonably closely matched to the component lead diameter for good soldering, an allowance of 0.15 to 0.3mm greater than the lead diameter giving the best results. Automatic insertion machines may require larger allowances. There is a trade-off between the number of differently-sized holes and board cost, so it is not advisable to specify different hole diameters for each different lead diameter. Typically you will use 0.8mm diameter for DIL package leads and most small components, 1.0mm for larger components and other sizes as required. Remember to specify hole diameters *after* plating on PTH boards. Also, remember to double-check component lead diameters against individual holes, when you're writing up the board specification - or, if your CAD system marks up hole diameters for you, make sure its component library is up-to-date. Some capacitors and power rectifier diodes in particular have larger leads than you may think! It is embarrassing to have to tell your production department to drill out holes on a double-sided PTH board, and then solder both sides, because you got a lead diameter wrong. It is even more embarrassing if it's a multilayer board, because you cannot drill out holes that connect to internal layers.

Vias

Via holes – plated-through-holes which join tracks on opposite sides of the board – can be any reasonable diameter, subject to current rating. The smallest useable diameter is linked to board thickness, and generally a ratio of thickness to diameter of 4:1 will not present too many problems in plating. But, unless you are forced into small diameter vias by constraints of packing density, you should keep them either equivalent to the smallest component hole diameter so that drill sizes are kept to a minimum, or make them one size smaller (e.g. 0.6mm) so that the likelihood of false insertion of component leads is low.

Pads

Pads can be oval or round. Oval pads for dual-in-line packages on a 0.1" pitch are a hangover from early days in pcb technology when a large pad area was needed to assure a good soldered joint and good adhesion of the pad to the board. This is still advisable for non-PTH boards, and a typical round pad for a 0.8mm diameter hole would be around 2mm diameter, which leaves no room for tracking between pins. An oval allows one track between each pad. Some care is needed with oval pads as the hole-to-pad tolerance in the width dimension can be tight.

PTH technology reinforces the pad-to-board bond on both sides by the plating in the barrel of the hole, and the solder flows into the annulus between lead and hole plating, so that large or oval pads are unnecessary on PTH boards. The pad need only contain the drilled hole before plating, with allowance for all manufacturing tolerances. For a 0.8mm diameter hole the pad diameter can be between 1.3 and 1.5mm (Figure 2.3).

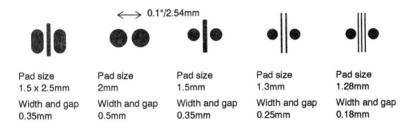

	0.1"/2.54mm			
Pad size 1.5 x 2.5mm	Pad size 2mm	Pad size 1.5mm	Pad size 1.3mm	Pad size 1.28mm
Width and gap 0.35mm	Width and gap 0.5mm	Width and gap 0.35mm	Width and gap 0.25mm	Width and gap 0.18mm

Figure 2.3 Spacing and dimensions for tracks between IC pads

Pads for larger holes on non-PTH boards should exceed the hole diameter by at least 1mm in order to obtain good adhesion. The ratio of pad-to-hole diameter should be around 2 for epoxy-glass and 2.5 - 3 for phenolic paper boards.

2.2.3   Track routing

The first rule of good routing is to minimise track length. Short tracks are less susceptible to interference and crosstalk, have lower parasitic reactances and radiate less. Routing should proceed interactively with package placement to achieve this. It is often possible to improve routing prospects when using multi-function packages (typically gates or op-amps) by swapping pins, so unless there are over-riding circuit considerations you shouldn't fix pin-outs on these packages until the layout has been finalised. A good CAD package with an extensive and intelligent component library will do this automatically. Similarly, you may have decided on grounds of package economy to use up all the functions in a package, but this could be at the expense of forcing long tracks to one or other function. An optimum board layout may require a few more packages than the minimum.

Many CAD auto-routing software packages will lay all tracks of one orientation on one side of the board and all of the other orientation on the other side. This works and is fast, especially for low-performance digital boards, but hardly ever produces an optimum layout in terms of minimum track length and number of via holes, and can be disastrous on analogue boards. Normally you should anticipate expending some skilled design effort at the layout stage in cleaning up the CAD output to produce a cheaper board with better electrical performance.

Track grid

To apply some discipline to the routing in the early stages, it is best to run tracks on a grid that aligns with the component holes, which themselves should be on a larger grid – usually 0.1"/2.54mm spacing, which is the universal standard for component pin-outs. This allows for easy and accurate automated drilling and component insertion. The track grid can be 0.05"/1.27mm but in the later stages you may have to depart from this for optimum tracking, and it should not be regarded as inviolate. Tracks should not be run closer than 0.5mm from the edge of the board.

45° angled bends are preferable to right-angles, as they allow a slight increase in tracking density. Right angles and acute angles in tracks are best avoided as they provide opportunities for etchant traps and the subsequent risk of track corrosion. When two tracks meet at an acute angle the join should be filleted to prevent this.

From a mechanical point of view, the aim should be to balance the total coverage of copper on both sides of a double-sided board, or on all layers of a multilayer. This guards against the risk of board warping due to differential strains, either because of thermal expansion in use or because of stress relief when the board is etched, and also assists plating.

2.2.4   Ground and power distribution

Much of what could be said here has already been said in Chapter 1. Layout of ground interconnections is as important on pcbs as it is between them. Common impedances should be avoided in critical parts of the circuit, but it is usually necessary to compromise with a ground "bus" for much of the circuit. This is perfectly acceptable at low frequencies, low gains, low currents and high signal levels, where the magnitude of voltages developed along the ground rail are well below circuit operating voltages.

Figure 2.1 shows resistance versus track width and this can be used to calculate dc voltage drops where necessary. 1mm track widths will suffice for most purposes, but it is a good idea to increase widths if space allows. The inductance of ground and power connections becomes important when high-frequency currents are being carried; the voltage drops now depend on rate of change of current.

Ground rail inductance

PC track inductance is primarily a function of length rather than width. The inductance of a track on its own is often misleading as it will be modified by proximity to other tracks, particularly those carrying return current. Rather than widen a track of given length to attempt to reduce its inductance, a better solution is to split it into several narrow tracks as shown in Figure 2.4.

too much inductance   not much improvement   greater improvement

Figure 2.4 Track inductance

Three narrow tracks will not give a threefold reduction because of mutual inductance effects, but it will be an improvement over a single thick track. Wider spacing relative to track width gives better results. A much more effective way of reducing total power and ground inductance is to run the outward and return paths very close together. This maximizes their mutual inductance and since the currents now oppose each other, the total inductance is *reduced* by the mutual inductance rather than increased. One way to achieve this is to run power and ground rails with identical geometry on opposite sides of the board. But, if a compromise is necessary it is better to use the board space to produce the best ground system possible and control power rail noise by decoupling (see section 6.1.4).

Gridded ground layout

Good low-inductance power and ground rails can be assured on double-sided and multilayer boards by either of two techniques: grid layout, or an overall or partial ground plane. The former is really an approximation of the latter. Some authorities suggest that a grid (Figure 2.5) is just as effective as a ground plane. For many purposes, particularly when the layout is a regular pattern of DIL ICs, and if the power and ground grids are well decoupled, this is true. Certainly a well-designed grid will be more effective than a poorly designed ground plane. A grid allows you to minimise sections

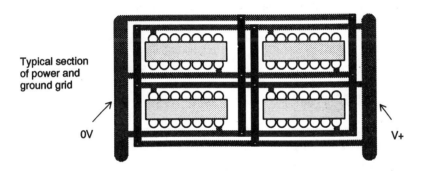

Typical section
of power and
ground grid

0V   V+

Figure 2.5 Gridded ground and power rail layout

of common impedance and ensures that lengthy ground separations are interconnected by several current paths. It is not advisable for sensitive analogue layouts when you need control of ground return current paths. In any case, an ideal grid will be too restricting for the signal paths and some compromise will be needed.

The ground plane

A ground plane scores when, as in analogue circuits or digital circuits with a mixture of package sizes, ground connections are made in a random rather than regular fashion across the layout. A ground plane is not necessarily created simply by filling all empty space with copper and connecting it to ground. There should be a minimum of interruptions to it, and for this reason ground planes are rather more successful on a multilayer board where an entire layer can be devoted, one to the power plane and one to the ground. This has the additional advantage of offering a low impedance between power and ground at high frequencies because of the distributed inter-plane capacitance. It is not always easy to make the decision between a four-layer board with internal power and ground planes, or a dense double-sided with gridded power and ground distribution. Generally the required package density is the deciding factor. Experience is the only real guide.

Individual holes make no difference to the ground plane, but large slots (Figure 2.6(a)) do. When the ground plane is interrupted by other tracks or holes the normal low-inductance current flow is diverted round the obstacle and the inductance is effectively increased. Interruptions should only be tolerated if they don't cut across lines of high dI/dt flow. Even a very narrow track interconnecting two segments of ground plane is better than none. At high frequencies, and this includes digital logic edge transitions, current tends to follow the path that encloses the least magnetic flux, and this means that the ground plane return current will prefer to concentrate under its corresponding signal track.

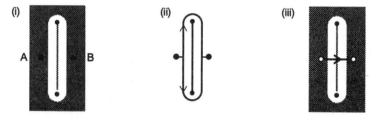

(a) When current flows from A to B, (i) is equivalent to (ii). (iii) is preferable

(b) Connecting pads to the ground plane

Figure 2.6 Ground plane connections

Some board manufacturers do not recommend leaving large areas of copper because it may lead to board warping or crazing of the solder resist. If this is likely to be a problem, you can replace solid ground plane by a cross-hatched pattern without seriously degrading its effectiveness. To make a connection to the ground plane – or any other large area of copper – you should "break out" the pad from the area and connect via one or more short lengths of narrow track (Figure 2.6(b)). This prevents the plane acting as a heatsink during soldering and makes for more reliable joints.

Proper, full analysis of power and ground current distribution is far too complex for most applications and it should be sufficient to understand the mechanisms involved so that you can avoid the most serious mistakes.

2.2.5   Package placement

There is a mix of electrical and mechanical factors to consider when placing components and ICs. Normally the foremost is to ensure short, direct tracks between components and it is always worth interactively optimising the placement and track routing to achieve this. You may also face thermal constraints, for instance precision components should not be next to ones which dissipate power, or you may have to worry about heat removal.

Over and above individual requirements there is also the general requirement of producibility. There should be continuous feedback between you as circuit designer and the production and service departments to make sure that products are easy, and therefore cheap, to produce and repair. Component and package placement rules should evolve with the capabilities of the production department. Some examples of producibility requirements follow:

- auto-insertion machines work best when components and packages are all facing the same way and are positioned on a well-defined grid

- small tubular components (resistors, capacitors and diodes) should conform to a single lead pitch to minimise the required tooling heads. It doesn't matter what it is (0.4" and 0.5" are popular) as long as it's constant

- inspection is easier if all ICs are placed in the same orientation, i.e. with each pin 1 facing the same corner of the board, and similarly all polarised components are facing the same way

- spacing between components should take into account the need to get test probes and auto-insertion guides around each component

- spacing of components from the board edges depends on handling and wave soldering machinery which may require a clear area on at least two edges

- if the board is to be wave soldered, rows of adjacent pins are best oriented across the direction of flow, parallel to the wave, to reduce the risk of solder bridges between pins or pads

2.2.6   Component identification

Most pcbs will have a legend, usually yellow or white, screen-printed on the component side to describe the position of the components. This can be useful if assembly of the finished unit relies on manual insertion, but its major purpose is for the test and service departments to assist them in seeing how the board relates to its circuit diagram. With low-to-medium density boards you can normally find space beside each component to indicate its number, but as boards become more densely packed this becomes increasingly difficult. Printing a component's ID underneath it is of no use to the service engineer, and if component placement is automatic it's no use to the production department either. Perticularly if the board consists almost entirely of DIL ICs, these can be identified by a grid reference system with the grid co-ordinates included on the component side track artwork. Therefore you should consider whether the extra expense of printing the legend onto a high-density board is justified.

Assuming it is, there are a few points to bear in mind when creating the legend master. If possible, print onto flat surfaces, not over edges of tracks or pads; the uneven surface tends to blur the print quality. Never print ink over or near a hole (allowing for tolerances). Even if it's not a solderable hole, the un-printed ink will build up on the screen and after several passes will leave a blot, ruining the readability. If there are large numbers of vias, these should be filled in or "tented" to prevent the holes remaining and rendering areas of the board useless for legend printing.

Polarity indication

There are several ways of indicating component polarity on the legend. Really the only criterion for these should be legibility *with the component in place*, so that they are as useful to inspectors and test engineers as they are to the assembly operators. Once a particular method has gained acceptance, it should be adhered to; using different polarity indicators on different boards (or even on the same board) is a sure way of confusing production staff and gaining faulty boards.

2.2.7   Solder resist

Also known as solder mask, the solder resist is a thin, tough coating of insulating material applied to the board after all copper processing has been completed. Holes are left in the resist where pads are to be soldered. It serves to prevent the risk of short circuits between tracks and pads during soldering and subsequently, and is also sometimes used as an anti-corrosion coating and to provide a dark, uniform background for the component identification legend. It can be a screen printed and oven cured epoxy resin, a photographically exposed and developed dry film or a photo-cured liquid film.

Screen printed resists

Screen printed epoxy resin is a well-established method and is inexpensive, but achievable accuracies are poor compared to modern etching accuracies. The consequence of this is that there has to be an allowance for mis-registration and resist bleeding of about 0.3 - 0.4mm between the edges of pads and the edge of the solder resist pattern. It is easy to generate the artwork for this - simply repeat the pad pattern with oversize pads, and generate a negative photographic image - but it cuts into the spacing between pads and if fine tracks are run between pads they may not be completely covered by the resist. Figure 2.7 shows this effect. This nullifies the supposed purpose of the resist, to prevent bridges between pads and adjacent tracks! Also, screen printed resists over large areas of copper that has been finished with tin-lead plating may crack when the board is wave soldered, as the plating melts and reflows. This is unsightly but not normally dangerous, as long as you do not rely upon the resist as the only corrosion barrier.

track exposed by poor registration

Figure 2.7 Mis-registration of the solder resist

Dry film

Dry film resists are capable of much higher registration accuracy and resolution (typically 0.1mm) and are therefore preferred for high-density boards. They have their own problems, apart from expense, the main one being lack of adhesion to poorly-prepared board surfaces.

A solder resist should not be regarded as essential. It can be useful in reducing the risk of board failure through surface contamination or solder bridges, but is not infallible. There is a danger that it is specified without thought, or used as a crutch to overcome bad soldering practices. A well-designed board in a good production environment may well be able to do without it.

2.2.8   Terminations and connections

Any pcb that is part of a system must have connections to it. In the simplest case this is a wire soldered to a pad. If the board is plated-through then this approach is acceptable, as the combined strength of solder in the plated hole plus the pad lands on both sides will be enough to cope with any normal wire flexing. Wires should not be soldered straight to non-PTH boards because the wire strains will be taken by the pad-to-board bond, which will quickly fail. Greater mechanical strength is needed.

This can be offered by feeding the wire back through a second hole in the board to give a measure of strain relief. Alternatively, use staked pins or "fish-beads". A pin press-fitted into a hole transmits all mechanical strains directly into the board laminate, so the reliability of the pad-to-track transition is unaffected. Wire dressing to the pin is slightly more labour-intensive but this is not usually a disadvantage. Fish-beads are easier to use but more expensive than pins, and because they are not staked to the board they can transmit some strain to the pad. Both types are eminently suitable for individual test points. Figure 2.8 shows the options.

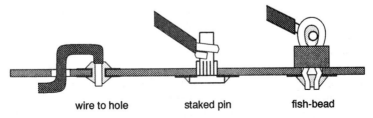

wire to hole   staked pin   fish-bead

Figure 2.8 Wire-to-board connections

Direct wire connections should not be dismissed out of hand because they are, after all, cheap. But as soon as several connections must be taken to the board, or repetitive disconnection is required, you will automatically expect to use a multi-way pcb connector. This can take one of two forms, a two-part moulded male/female system, or an "edge" connector.

Two-part connectors

There are many standard pcb connectors available and it is impossible to do justice to all of them. Popular ones are the DIN-41612 range for Eurocard-sized module boards, the many variants of square-pin stacked connectors pioneered by Molex and available from several sources in pin pitches from 0.05" to 0.2", the insulation displacement (IDC) types with pc-mounting headers and free sockets, and the subminiature "D" range to MIL-C-24308 for external data links. All you can do is to compare data sheets from several manufacturers to find the best overall fit for your particular requirement. In connectors it is especially true that you get what you pay for in terms of quality.

Contact resistance will be important if the connector carries appreciable current, as will be the case for power supply rails. Even more important is whether a low contact resistance will be maintained over time in the face of corrosion and repeated mating cycles. This depends mainly on the thickness of gold plating on the mating surfaces. A wise precaution is to dedicate several ways of a multi-way connector in parallel for each power and ground rail, to guard against the effects of increasing contact resistance and faulty pins.

Insertion and withdrawal force specifications tend to be overlooked, but serious damage to a board can result if too much force has to be used to plug it in. Conversely, if the withdrawal force is low and/or the plug-socket pair is not latched, there is the risk of the connectors falling apart or vibrating loose. As with single wires, it is better to provide a separate route for diverting the mechanical strain of the interconnection into the board laminate rather than relying on the solder joints to individual pins. Connector mouldings which allow for nut-and-bolt fixing are best in this respect. Incidentally, remember to specify on the production drawing that fixings should be tightened *before* the pins are soldered - otherwise strain will be put on all the soldered joints as soon as the nuts are torqued up, resulting in unreliable connections later.

Edge connectors

Edge connector systems are popular with many designers because of their relative cheapness. In this case, the connector "finger" pattern is part of the pcb layout at the board edge, and the board itself plugs into a single-piece female receptacle on the mother-board or chassis. The board should have a machined or punched slot at some point in the pattern which mates with a blanking key in the receptacle, to ensure that the assembly is plugged in the right way round and to align the individual connectors. The ends of the receptacle can then be left open. This is safer than aligning the board edge with the end of a closed receptacle, as it is more accurate and less susceptible to damage.

The fingers on the board must be protected from corrosion by gold plating. The plating should cover the sides as well as the tops of the fingers, otherwise long-term corrosion from the edges will be a problem. Pay attention to dimensional tolerancing of the pcb, both on board thickness to ensure correct contact pressure and on machining to ensure accurate mating. A useful trick if you have spare edge connector contacts is to run them to a dummy pad in some unused space inboard. They will prove invaluable when you discover a need for more connections during prototype testing!

One final point on pcb connectors: make sure that your chosen connector type and board technology are compatible. Very high-density connectors are available which have multiple rows of pins at less than 0.1" spacing. These require very fine tracking to get between the outer rows to the inner ones. Also, they are a nightmare to assemble to the board, and you may find that your production department want larger holes than you bargained for, which makes the tracks even finer. Don't go for high-density connectors unless you really have to: stick with the chunky ones.

2.3   Surface protection

The insulation resistance between adjacent conductors on a bare pcb surface depends on the conductor configuration, the surface resistance of the base material, processing of the board and environmental conditions, particularly temperature, humidity and contamination. For a new board with no surface contamination the expected insulation resistance between two parallel conductors can be derived from

$$R_i \quad = \quad 160 \cdot R_m \cdot (w/l)$$

where R_m is the material's surface resistance specification at a given temperature (> 50,000MΩ for epoxy glass, > 1000MΩ for phenolic paper)
               w is the track spacing
               l is the length of the parallel conductors

Variations in surface resistance

Generally this value is of the order of thousands of megohms and could be safely ignored for most circuits. Unfortunately it is not likely to be the actual value you would measure. This is because plating and soldering processes, dust and other surface contamination, moisture absorption and temperature variation will all have the effect of lowering the insulation resistance. Variations between 10 and 1000 times less can be observed under normal operating conditions, and a severe environment can reduce it much more.

When you are working with high impedance or precision circuits you may not be able to ignore the change in surface resistance. Until you realise what the problem is, its effects can be mysterious and hard to pin down: they include variability of circuit parameters with time of day, handling of the board, weather (relative humidity), location and orientation of the assembly, and other such factors that normally you would expect to be irrelevant. Typical variations are change of bias point on high input impedance amplifiers and unreliable timing on long time constant integrators.

Circuit design vs. surface resistance

There are a number of strategies you can use to combat the effects of surface resistance. The first and most obvious is to keep all circuit operating impedances as low as possible, so that the influence of an unstable parallel resistance in the megohm region is minimal. In some cases – micropower circuits and transducer inputs for example – you don't have the option. In other cases a change in circuit philosophy could be beneficial; a long time constant analogue integrator or sample-and-hold might be replaced by its digital equivalent, with an improvement in accuracy and repeatability.

If there is a particular circuit node which must be maintained at a high impedance, the wiring to this can be kept off the board by taking it to a PTFE stand-off insulator. PTFE has excellent surface resistance properties even in the presence of contamination. Alternatively, reducing the length of high impedance PC tracks and increasing their distance from other tracks may offer enough improvement, at the expense of packing density. Simply re-routing an offending track might help: if a power rail is run past a high impedance node which is biased near 0V, you have an unwanted potential divider which will pull the bias voltage up by an unpredictable amount. Put the power rail track elsewhere.

2.3.1   Guarding

The next level of defence is guarding. This technique accepts that there will be some degree of leakage to the high impedance node, but minimizes the current flow to it by surrounding it with a conductive trace that is connected to a low impedance point at the same potential. Like the similar circuit technique of bootstrapping, if the voltage

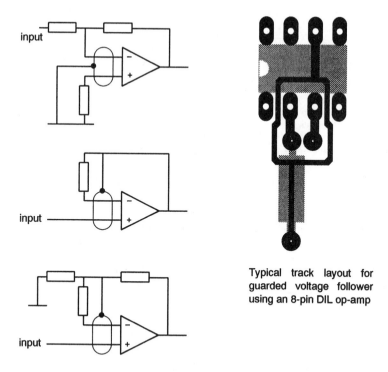

Typical track layout for
guarded voltage follower
using an 8-pin DIL op-amp

Figure 2.9 Input guarding for various op-amp configurations

difference between two nodes is forced to be very low, the apparent resistance between them is magnified. The electrical connections and pc layouts for the guards for the basic op-amp input configurations are shown in Figure 2.9. The guard effectively absorbs the leakage from other tracks, reducing that reaching the high impedance point.

There should be a guard on both sides of a double-sided board. Although the guard virtually eliminates surface leakage, it has less effect on bulk leakage through the board; fortunately this is orders of magnitude higher than leakage due to surface contamination. The width of the guard track is unimportant as far as surface resistance goes, but a wider guard will improve the effect on bulk resistance.

Guarding is a very useful technique but obviously requires some extra thought in circuit design. Generally, you should consider it in the early design stages whenever you are working with impedance levels which are susceptible to surface resistance variations. You may then be in the happy position of never noticing the problem.

2.3.2   Conformal coating

If none of the above methods are sufficient, or if they are inapplicable, or if the working environment of the board is severe (relative humidity approaching 100%, conductive or organic contamination present, corrosive atmosphere) then you will have to go for conformal coating. This is *not* a decision to be taken lightly; try everything else first. Coating adds pain, sweat and expense to the production process and the following discussion will outline why.

Coating vs. encapsulation

Note that conformal coating is not the same as encapsulation, or potting, which is even less desirable from a production point of view. Encapsulation fills the entire unit with solid compound so that the end result looks rather like a brick, and is used to prevent third parties from discovering how the circuit works, or to meet safety approvals, or for environmental or mechanical protection. A badly-potted unit will probably work like a brick too: differential thermal expansion as the resin cures can crack poorly-anchored pc tracks, and the faulty result is unrepairable. Conformal coating covers the board with thin coats of a clear resin so that the board outline and components are still visible. It provides environmental protection only. Occasionally it may help in meeting safety approval clearances but this is rare.

The main environmental hazard against which conformal coating protects is humidity. The popular coating types, acrylics, polyurethanes, epoxy resins and silicones are all moisture-proof. Most offer protection against the common chemical contaminants: fluxes, release agents, solvents, metal particles, finger grease, food and cosmetics, salt spray, dust, fuels and acids. Acrylics are rather less resistant to chemical attack than the others. A conformal coating will not allow closer track spacing when this is determined by the surface insulation properties of the board, but it does eliminate degradation of these properties by environmental factors: *if it is used properly.*

Steps to take before coating

The first point to remember is that a conformal coating seals in as well as out. The cleanliness of the board and its low moisture content are paramount. If residual contaminants are left under the film, corrosion and degradation will continue and will eventually render the coating useless. A minimum of three steps must be followed immediately prior to coating:

- vapour degrease in a solvent bath (note that with increasing concern about environmental pollution and ozone depletion, traditional cleaning fluids, particularly CFCs, are being phased out)

- rinse in deionised water or ethyl/isopropyl alcohol to dissolve inorganic salts

- oven bake for 2 hours at 65 - 70°C (higher if the components will allow it) to remove any residual solvent and moisture.

After cleaning and baking, the assembly should only be handled with rubber or lint-free gloves. If the cleaned boards are left for any appreciable time before coating, they will start to re-absorb atmospheric moisture, so they should be packed in sealed bags with dessiccant.

Application

The coating can be applied by dipping or spraying. The application process must be carefully controlled to produce a uniform, complete coat. Viscosity and rate of application are both critical. At least two and preferably three coats should be applied, air- or oven-drying between each, to guard against pinholes in each coat. Pot-life of the coating material – the length of time it is useable before curing sets in – is a critical parameter since it determines the economics of the application process. Single-component solvent based systems are preferable to two-part resins in this respect and also because they do not need metering and mixing, a frequent source of operator error. On the other hand, solvent based systems require greater precautions against operator health hazards and flammability.

Nearly all boards will require breaks in the coating, for connectors or for access to trimming components and controls. Aside from the question of the environmental vulnerability of these unprotected areas, such openings require masking before the board is coated, and removal of the mask afterwards. Manual application of masking tape, or semi-automatic application of thixotropic latex-based masking compound, are two ways to achieve this.

Test and rework

Finally, once the board is coated, there is the difficult problem of test, rework and repair. By its nature the coating denies access to test probes, so all production testing must be done before the coating stage. Acrylics and polyurethanes can be soldered through or dissolved away to achieve a limited degree of rework, but other types cannot. After rework, the damaged area must be cleaned, dried and re-coated to achieve a proper seal. Ease of rework and ease of application are often the most important considerations in choosing a particular type of coating material.

You might now appreciate why conformal coating is never welcome in the production department. Because it is labour-intensive it can easily double the overall production cost of a given assembly. Specify it only after long and careful consideration of the alternatives.

2.4   Surface mount

Rather than mounting components by their leads through holes in the board, it is also possible to mount them on one or both surfaces of the board, held in place only by solder between pads on the board and the component terminations. Lead holes are unnecessary, components are smaller and the net result is a considerable increase in component density per unit area. Board assemblies made in this manner are known as surface mount assemblies (Figure 2.10).

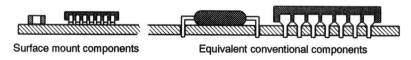

Surface mount components   Equivalent conventional components

Figure 2.10 The difference between surface mount and conventional

History

In the latter half of the 1980s, surface mount technology (SM, or SMT) suddenly became fashionable. This was due to a combination of market pressure for smaller, more complex electronic assemblies, increasing sophistication of the pcb manufacturers and the achievement of critical mass in the supply of SM components. Quite probably it was also a response to the market dominance of the Japanese in the consumer sector, who learnt to apply SM techniques successfully to large-volume products years before the rest of the industry. Wild predictions were made about the eventual replacement of all other methods of construction by surface mounting. In the event, the euphoria has subsided, and SM is now seen as an important technique for its appropriate applications, complementary to conventional through-hole mounting methods.

A great deal has been written about the detail of SM design and production, so this section will only touch on the more general points, to help in the decision between one or other method. The advantages and disadvantages of SM construction versus conventional can be summarized as follows:

Advantages

- *size*. Very much higher packing densities can be achieved. Components can be mounted on both sides if necessary. Applications that were previously unachievable are now possible.

- *automation*. SM component placement and processing lends itself to fully-automated assembly and so it is well suited to high-volume production. Unit assembly costs can be lowered.

- *electrical performance*. Reduced size leads to higher circuit speeds and/or lower interference susceptibility. Higher performance circuits can be built in smaller packages – a prime marketing requirement.

Disadvantages

- *investment*. To properly realise the potential of automated production, a sizeable capital investment in machinery is needed. This is normally measured in hundreds of thousands of pounds and is beyond the means of smaller companies.

- *experience*. A company cannot jump into SM production overnight. At every stage in the design and production process, new skills and techniques have to be learnt.

- *components*. Although there has been an explosion in SM component availability in the last five years, still there are specialized types which are only obtainable as through-hole mounting. Higher-power and large components can never be surface mounted. Even among common devices, standardization is slow in coming.

- *criticality of mechanical parameters*. Traditionally, any electrically equivalent component would do if it fitted the footprint. With SM, mechanical equivalence is as critical as electrical equivalence, because the placement and soldering processes are that much less forgiving. This leads either to unreliability or to difficulties with component sourcing. Also, solderability and component shelf life become dominant issues.

- *test, repair and rework*. It is possible to test and rework faulty SM boards, with tweezers, a hot-air gun and a magnifying glass. It is a lot easier to test and rework conventional boards. Be prepared for extra expense and training at the back end of the production line.

The investment that is needed for a company that is contemplating doing its own SM production extends beyond simply acquiring the production equipment. There is probably an equivalent investment in new CAD design tools, new storage and procurement systems (to cut down component shelf life), new test equipment and rework stations, plus a hefty amount of time for re-training. Many companies will therefore prefer to use the services of a sub-contract assembly house, despite a loss in their own profit margin, if their product volume cannot justify this investment. An important benefit of this approach is that it allows a firm to experiment with SM, and

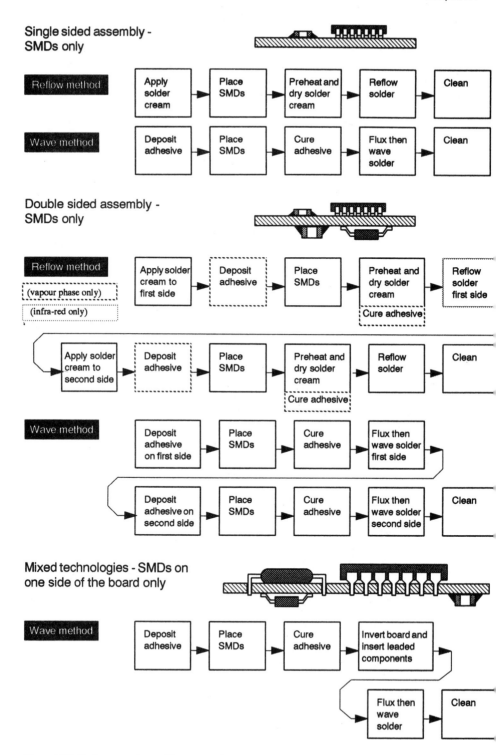

Figure 2.11 Surface mount assembly process stages

gain some market and product experience with it, before full commitment. The other side of the coin is that the production staff do not gain any significant experience.

The stages of design, assembly and test are very much more tightly coupled in SM than they need to be in conventional manufacture. The successful production and testing of an assembly is critically related to the pcb layout and design rules employed.

2.4.1   Surface mount design rules

Pad dimensions and spacing are relative to the component body and leadouts and depend on the soldering method that will be used with the board. Components can either be placed over a dot of adhesive to hold them in place and then run through a wave-soldering operation, or they can be placed on a board which has had solder paste screen printed onto the pads, in which case the tackiness of the solder paste holds them lightly in position until it is reflowed, either by an infra-red oven or a hot vapour bath. Figure 2.11 shows the different stages for each type of production.

Solder process

If the board will be wave-soldered, then the IC packages should be oriented along the direction of board travel, across the wave, and a minimum spacing should be observed. This optimises solder pick-up and joint quality. Pad dimensions need to be fairly large in order to take up placement tolerances, since the absolute position of the package cannot be altered once the adhesive is cured. The advantage of wave soldering is that it can simultaneously process surface mount and conventional components, if they are mounted on opposite sides. The maximum height of the SM components will be determined by the risk of being detached from the board when passing through the wave.

Vapour phase or infra-red soldering allows tighter packing and smaller pads, and the orientation is not critical. When the solder paste reflows, surface tension will draw misaligned components into line with their pads, so that placement tolerance is less critical. Shadowing can be a problem with components of varying height placed close together, as can variable heat absorption if there are large and small packages with different heat reflection coefficients. Board layout must start from the knowledge of which soldering method will be used, and also what tolerances will be encountered in the placement process and in the components themselves. If wave-soldering pad dimensions are used for reflow soldering, solder joint quality will suffer. Most layout designers use different pad dimensions for the two systems.

Printed circuit board quality

PCB finish is more important than it is with conventional components. The overriding requirement is flatness of the surface, since component sizes are that much smaller and since good soldered joints depend on close contact of the leadouts with the pads. A solder resist is essential to control the soldering process. Dry film resists are to be preferred to screen printed (see section 2.2.7) since their thickness is well controlled; also because the tolerances on solder mask windows are that much tighter. There must be no bleeding of the resist onto the pads. Hot air levelling of the tin/lead finish on the solder pads is desirable to prevent the bumps that form on the surface of ordinary reflowed tin/lead.

Thermal stresses

Differential thermal expansion is a potential reliability threat for some SM components.

Chip resistors and capacitors, and leadless chip carriers (LCCs), are made with a ceramic base material whose coefficient of thermal expansion is not well matched to epoxy fibreglass. Originally these components were developed for hybrid circuits which use ceramic substrates and good thermal matching is possible. At the same time, leadouts for these components are deposited directly on the ceramic so that there is no compliance between the leadout (at the soldered joint) and the case. As a result, strains set up under thermal cycling can crack the component itself or the track to which it is soldered.

You should not therefore use the larger ceramic or LCC components directly on epoxy fibreglass board. Leaded SM components such as small outline or flat-pack ICs do not suffer from this problem because there is a section of compliant lead between the soldered joint and the package. Plastic leaded chip carriers (PLCCs) with J-lead construction are useable for the same reason, and small chip ceramic components are also useable, because of their small size.

Cleaning and testing

Cleaning an SM board is trickier than for a conventional assembly because there is less of a gap underneath the packages. Flux contamination can get into the gaps but it is harder to flush out with conventional cleaning processes. There is considerable effort being put into developing solder fluxes that do not need to be cleaned off afterwards.

Testability is an important consideration. It is bad practice to position test probes directly over component leadouts. Apart from the risk of component damage, the pressure of the probe could cause a faulty joint to appear sound. All test nodes should be brought out to separate test pads which have no component connections to them, and which should be on the opposite side of the board to the components. There is an extra board space overhead for these pads, but they need be no more than 1mm in diameter. Testing a double-sided densely packed SM assembly is a nightmare.

2.5 Sourcing boards and artwork

Before we leave the subject of pcbs, a short discussion of board procurement is in order. There are two stages involved: generation of the artwork and associated documents, and production of the boards. The two are traditionally separated because different firms specialize in each, so that having generated the artwork from one source you would take it to another to have the boards made. Some larger pcb firms have both stages in-house, or have associated operations for each.

2.5.1 Artwork

The artwork includes the track and solder resist patterns, the hole drilling diagram, the component legend and a dimensional drawing for the board. The patterns are generated photographically, either direct from CAD output or by photo-reduction from a scaled-up master artwork. To create the artwork you can either do it yourself in-house using manual tape and film or a CAD system, or you can take the work to a bureau that specializes in PC artwork and has its own CAD system. There are advantages and trade-offs both ways.

Let's assume that your company has drawing office facilities for both manual taping and CAD. Which of these you choose will depend on the complexity of your design, availability and suitability of resources and any laid-down company policy. The more complex the design, the greater incentive there is to use CAD, but there is also a good argument for using CAD for all designs, to ensure a consistent output or simply to keep in practice. But it may also be that your company isn't very rich and cannot afford a major CAD suite that is under-utilised, so that the in-house CAD system is not available when you want it, or doesn't offer the facilities you need for a new design.

Using a bureau

So you may be faced with the choice of going to an outside bureau. Reasons for doing this are if

- your own company doesn't have any artwork facilities at all;
- you do have the facilities but they are insufficient for or inappropriate to your design;
- you have all the facilities you need but they aren't available in the timescale required.

There are less tangible advantages of a bureau. Its staff must by nature all be highly skilled and practised layout draughtsmen, and they should be using reasonably up-to-date and reliable CAD systems, so the probability of getting a well-designed board out in a short time is often higher than if you did it in-house. They will also be able to give you guidance in the more esoteric aspects of pcb design that you may not be familiar with. This can be particularly important for new surface mount designs, especially if you have not yet developed your own SM design rules. A bureau that is interested in keeping its customers will be flexible with its timescales, and may well offer a faster turn-round than you could achieve in-house. Even the cost should be no more than you would have to pay in-house, since the bureau is using its resources more efficiently. Finally, a bureau that works closely with a board manufacturer will be able to generate artwork that is well matched to his requirements, thereby freeing you of an extra and possibly unfamiliar specification burden.

Disadvantages of a bureau

There are three possible disadvantages. One is confidentiality: if your designs go outside, there is always the possibility that they will be compromised, whatever assurances the bureau may give. Another is that you are then locked in to that bureau for future design changes, unless you happen to have the same CAD system (beware: different software versions can be incompatible) in-house. The risk is that the bureau might not be as responsive to requests for changes in the future as it would be to the first job. You have to balance this against what you know of the bureau's past history.

Finally, you may feel that you have less control over the eventual layout. This is not such a threat as it may seem. You always have the option to go and look over the layout designer's shoulder while they are doing the work; select a bureau nearby to avoid too much travel. But more importantly, giving your design to an outsider imposes the very beneficial discipline that you need to specify your requirements carefully and in detail. Thus you are forced to do more rigorous design work in the early stages, which in the long run can only help the project.

2.5.2   Boards

PCB suppliers are not thin on the ground. It should be possible to select one who can offer an acceptable mix of quality, turn-round and price. There is, though, a lot of work involved in doing this and once you have established a working relationship with a particular supplier it is best to stay with him, unless he lets you down badly. Any board manufacturer should be happy to show you round his facilities, and should be able to be specific about the manufacturing limits to which he can work. You may also derive some assurance from going to a quality-assessed supplier, or it may be company policy to do so. While market competition prevails, using a quality-assessed supplier should be no more expensive than not.

The major difference lies between sourcing prototype and production quantity boards. In the prototype stage, you will want a small quantity of boards, usually no more than two or three, and you will want them fast. A couple of weeks later you will have found the errors and you will want another two or three even faster, because the deadline is looming. The fastest turn-round offered on normal double-sided PTH boards is four or five days, and the cost for such a "prototype service" is between two and five times that for production quantities with several weeks' lead time.

Sometimes you can find a manufacturer who can offer both reasonable production and prototype services. Usually though, those who specialize in prototype service don't offer a good price on production runs, and vice versa. If you manage to extract a fast prototype service on the promise of production orders, you then have to explain to the purchasing department why they can't shop around for the lowest price later. This can be embarrassing; on the other hand they may be operating under other constraints, such as incoming quality. There are several tradeoffs that the purchasing department has to make which may not be obvious to you, and it pays to involve them throughout the design process.

Chapter 3

Passive Components

3.1 Resistors

Resistors are ubiquitous. Because of this their performance is taken for granted; provided they are operated within their power, voltage and environmental ratings this is reasonable, since after millions of accumulated resistor-years experience there is little left for their manufacturers to discover. But there are still applications where specifying and applying resistors needs to be handled with some care.

Let us start with an appreciation of the different varieties of resistor that are available. Table 3.1 (overleaf) is a guide to the common types that will be encountered in general circuit design. There are more esoteric types which are not covered.

3.1.1   Resistor types

Carbon

The most common resistor for commercial applications is the carbon film. It is definitely the cheapest – less than a penny in quantity. It also has the least impressive performance, apart from carbon composition, but it is normally adequate for general purpose use. The other type which uses carbon as the resistive element is carbon composition. Composition types have a ridiculously poor performance – they were the first type of carbon resistors ever made in quantity – but are still used in certain special applications.

Metal film

The next most common type is the metal film, in its various guises. This is nowadays the standard part for industrial and military purposes. The most popular varieties of metal film are hardly any more expensive than carbon film and, given their superior characteristics, particularly temperature coefficient, noise and power handling ability, many equipment manufacturers do not find it worthwhile to bother with carbon film.

Variants of the standard metal film cater for high resistance needs. (The "metal" in a metal film is a nickel-chromium alloy of varying composition for different resistance ranges.) The metal oxide resistor used to be more popular than it is now. At one time it offered a good performance/price tradeoff midway between carbon and metal film, but now that metal film prices have dropped it is primarily used for lower-resistance, medium power (0.5 – 5W) applications where wirewound is unsuitable.

Wirewound

For medium and high power (>2W) applications the wirewound resistor is almost universally used. It is fairly cheap and readily available. Its disadvantages are its bulk, though this allows a lower surface temperature for a given power dissipation; and that

Type	Ohmic Range	Power range	Tolerance	Tempco range	Manufacturers	Applications	Cost
Carbon film	2.2Ω-10M	0.25 - 2W	5%, 10%	-150 > -1000ppm/°C	Neohm, Dubilier, Rohm, Piher	General purpose/ commercial	< 1p
Carbon composition	2.2Ω-10M	0.25 - 1.0W	10%, 20%	+400 > -900ppm/°C	Neohm, Allen Bradley	Special	
Metal film (standard)	1Ω - 10M	0.125 - 2.5W	1%, 2%, 5%	+/-50 > 200ppm/°C	Philips, Neohm, Corning, Piher, Rohm	General purpose industrial & military	1 - 2p
Metal film (high ohm)	100K - 100M	0.5 - 1W	5%	+/-200 > 300ppm/°C	Philips	High voltage & special	5p
Metal glaze	1Ω-100M	0.25W	2%, 5%	+/-100 > 300 ppm/°C	Neohm	Small size	5p
Wirewound	0.1Ω-33K	2 - 20W (10W - 100W aluminium)	5%, 10%	+/-75 > 400ppm/°C	Philips, Welwyn, CGS, VTM, Erg	High Power	15 - 50p 50p - £1 (al)
Metal film (precision)	5Ω - 1M	0.125 - 0.4W	0.05 > 1%	+/-15 > 50ppm/°C	Philips, Resolute, Holsworthy, Beyschlag	Precision	10p - 50p
Wirewound (precision)	1Ω - 1M	0.1W - 0.5W	0.01 > 0.1%	+/-3 > 10ppm/°C	General Resistance, Vishay, Sfernice	Extra-precision	£2 - £20
Bulk metal (precision)	1Ω - 200K	0.33 - 1W	0.005 > 1%	+/-1 > 5ppm/°C	Vishay, Sfernice, Caddock, TDK	Extra-precision	
Resistor networks	100Ω - 100K (10Ω - 1M)	0.125 - 0.3W per element	2%	+/-100 > 300ppm/°C +/-50ppm/°C tracking	Bourns, Dubilier, CTS, Allen Bradley, Beckman	Multi-resistor	30 - 50p
Chip resistors	10Ω - 10M	0.1 - 0.5	2%, 5%	+/-100ppm/°C	Philips, Welwyn, Rohm, Murata, Kyocera	Surface mount, hybrids	1p - 2p

Notes: 1) This survey does not consider special types 2) Manufacturers quoted are those widely sourced in the UK at time of writing 3) Quoted ranges are for guidance only 4) Costs are typical for medium quantities

Table 3.1 Survey of resistor types

because of its construction it is noticeably inductive, which limits its use in high-frequency or pulse applications. Wirewound types are available either with a vitreous enamel or cement coating, or in an aluminium housing which can be mounted to a heatsink. Aluminium housings can offer power dissipations exceeding 100W per unit.

Precision resistors

Once circuit requirements start to call for accuracy and drift specifications exceeding the usual metal film abilities, the cost increases substantially. It is still possible to get metal film resistors of "precision" performance up to an order of magnitude better than the standard, though at prices an order of magnitude or more higher. Drift requirements of less than 10 parts-per-million per °C (ppm/°C) introduce many more significant factors into the performance equation, such as thermal emf, mechanical and thermal stress, and terminating resistance. These can be dealt with, and the resistive and substrate materials can be optimised, but the unit costs are now measured in pounds, and delivery times stretch to months.

Resistor networks and chips

Resistor networks are manufactured somewhat differently. A resistive ink is screen-printed onto a ceramic substrate to form many resistors at once, which is then encapsulated to form a multi-resistor single package. The resulting resistors have roughly the same performance as a low-grade metal film, though with reduced breakdown voltage and power handling ability.

Chip resistors use the same process to produce an individual part. The standard size of chip resistor (there are other sizes) is 3.2mm x 1.6mm x 0.6mm thick; they are intended for automatic placement on surface mount or hybrid assemblies. In fact, the resistive ink technique used for networks and chips can also produce standard axial-lead resistors (metal glaze) of small size, and can be used directly onto a substrate to generate printed resistors. This technique is frequently used in hybrid circuits and is very cost-effective especially when large numbers of similar values are required. It is possible to print resistors directly onto fibreglass printed circuit board, though the result is of very poor quality and cannot be used where a stable, predictable value (compared with conventional types) is required.

3.1.2   Tolerancing

Perhaps the most fundamental application question is that of accuracy and tolerancing. Process variations apply to all standard resistor manufacturers; for example, a single 68K resistor will not have an absolute value of 68,000 ohms. If it is a 5% part its value when supplied should lie between 64.6K and 71.4K (and occasionally outside, if the manufacturer's quality assurance isn't up to scratch). What are the consequences of this, for example, for the humble potential divider?

The unloaded value of V_O is not 5V. In the real circuit there are two worst-case values to take into account: when R1 is at the upper limit of its tolerance and R2 is at the lower limit, and vice versa. These two cases yield

$$V = 10 \times 64.6 / (64.6 + 71.4) = 4.75V \text{ or } -5\%, \text{ and}$$

$$V = 10 \times 71.4 / (71.4 + 64.6) = 5.25V \text{ or } +5\%$$

The general case is

$$V'/V = 1 - (2K \cdot R1 / [R2 \cdot (1 - K) + R1 \cdot (1 + K)])$$

where V is the output voltage with both resistors nominal, and V' is the output voltage when R1 is at its high tolerance K and R2 is at its low tolerance K (for 5% resistors, K = 0.05). If the resistors are equal, the voltage variation is the same as the resistor tolerance.

You have to check that neither of these cases lead to out-of-limits circuit operation; and to be thorough, this must be done for all critical resistor combinations in the whole circuit. The trick, of course, is to know which combinations are critical. In a complex resistor network it is not always easy to see which permutations produce worst-case results. In such cases a circuit simulator can quickly prove its worth.

Tolerance variations

It is sometimes tempting to hope that tolerance variations will cancel out over a medium-to-large sample, so that if a circuit parameter depends on several resistor values a sort of "average" tolerance which is less than the specified tolerance can be used. This is dangerously bad practice. Process similarities (the flip side of process variations) can often mean that a single batch of 5% resistors all have the same value to within, say, 1%, yet be 4% off their nominal value. This is quite a frequent occurrence, because the manufacturer may select parts for close tolerance purposes out of a batch of wider tolerance. What is left in the wide tolerance batch then has "holes" in the middle of the tolerance range. The manufacturer is perfectly entitled to ship them, but your test department will find a whole batch of assemblies built with perfectly good components and all showing the same fault.

If the standard 5% carbon film tolerance is not good enough, there is the option of the (slightly) more expensive 2% or 1% metal oxide or metal film types. Indeed, the price differential between these is now so small as a proportion of the overall cost of an

E6 ±20%	E12 ±10%	E24 ±5%	E48 ±2%	Additional E96 ±1%
1.0	1.0	1.0, 1.1	1.00, 1.05, 1.1, 1.15	1.02, 1.07, 1.13, 1.18
	1.2	1.2, 1.3	1.21, 1.27, 1.33, 1.40, 1.47	1.24, 1.30, 1.37, 1.43
1.5	1.5	1.5, 1.6	1.54, 1.62, 1.69, 1.78	1.50, 1.58, 1.65, 1.74
	1.8	1.8, 2.0	1.87, 1.96, 2.05, 2.15	1.82, 1.91, 2.00, 2.10
2.2	2.2	2.2, 2.4	2.26, 2.37, 2.49, 2.61	2.21, 2.32, 2.43, 2.55, 2.67
	2.7	2.7, 3.0	2.74, 2.87, 3.01, 3.16	2.80, 2.94, 3.09, 3.24
3.3	3.3	3.3, 3.6	3.32, 3.48, 3.65, 3.83	3.40, 3.57, 3.74
	3.9	3.9, 4.3	4.02, 4.22, 4.42, 4.64	3.92, 4.12, 4.32, 4.53
4.7	4.7	4.7, 5.1	4.87, 5.11, 5.36	4.75, 4.99, 5.23, 5.49
	5.6	5.6, 6.2	5.62, 5.90, 6.19, 6.49	5.76, 6.04, 6.34, 6.65
6.8	6.8	6.8, 7.5	6.81, 7.15, 7.50, 7.87	6.98, 7.32, 7.68, 8.06
	8.2	8.2, 9.1	8.25, 8.66, 9.09, 9.53	8.45, 8.87, 9.31, 9.76

Table 3.2 IEC63 standard component values

assembly that it is common to standardise on the 2% or 1% ranges for all applications, regardless of the actual requirement. Below 1% tolerance the field is taken over by precision resistors which are normally specified to have a particular required value, rather than being only available in the standard values. The standard value range specified by IEC63 is given in Table 3.2.

3.1.3   Temperature coefficient

Where precision resistors are needed, usually in measurement applications, another parameter becomes important: their resistive temperature coefficient (tempco), expressed in parts per million per °C (ppm/°C).

Standard metal film resistors have tempcos of the order of ±50 to ±200ppm/°C. The value of a 200ppm/°C part could change by up to 1% over a temperature range of 50°C. This does not mean that every resistor quoted at 200ppm/°C will change by this much, only that this is the maximum that you can expect. The actual temperature coefficient depends on the manufacturing process and also on the value. For instance, carbon film tempco varies from −150 to −1000ppm/°C depending on value. Precision wirewound, metal film and bulk metal resistors can achieve orders of magnitude better than this − 1ppm/°C is achievable − but are correspondingly expensive.

A typical use for precision resistors is in dividing down a stable voltage reference - for example, there is no point selecting a voltage reference with a tempco of 30ppm/°C and then dividing it down with 200ppm/°C resistors. The type used for R1 in this circuit is not

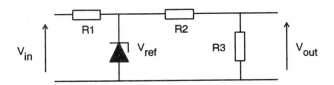

critical as input voltage regulation will usually be far more significant than value changes. On the other hand, if V_{out} is to be used as a reference derived from V_{ref}, then R2 and R3 must be of comparable stability to the reference voltage. Suppose that V_{out} is required to be 1.00V ±1.5% and to show 30 ppm/°C temperature stability. The reference is an LM385B-1.2 which has a specified voltage of 1.235V ±1% and an average temperature coefficient of 20ppm/°C.

To get 1.00V with nominal values and assuming R3 = 10K with no load current, then R2 will need to be 2.35K (not a standard value, though the closest in the E96 series is 2.37K). Taking worst-case tolerances for the LM385 and both resistors (V_{ref} high, R2 low, R3 high; V_{ref} low, R2 high, R3 low) and assuming both resistors have the same tolerance, then some calculation shows that the specified resistor tolerance should be better than 1.4%. Similarly the tempcos should be better than 26ppm/°C. These requirements would point to the use of 1% 25ppm metal film types.

Temperature changes can come either from ambient variations or from self-heating due to dissipated power. Any application which requires good resistance value stability should aim for minimum, or at least constant, power dissipation in the component. Manufacturers' data normally shows a graph of temperature rise versus power dissipation for a given resistor type and this should be checked if value stability has to be maintained through changes in dissipation.

3.1.4   Power

Power dissipation is, of course, one of the most important specifications a designer has to check for each component. Power dissipation results in temperature rise, which is determined by how fast the heat is conducted away from the body. The maximum body temperature usually occurs in the middle of the resistor, and this is known as the hot-spot temperature. Prolonged high temperatures (remembering that the quoted hot-spot temperature must be added to the maximum ambient) cause two things: a reduction in reliability, not only of the resistor but of components near it, and a resistance shift.

A good rule-of-thumb for a reliable circuit is never to allow a dissipation of more than half the rated power in each component; many companies have their own in-house rules based on experience or customer's specifications.

Power should be calculated under worst-case operating conditions: for example a resistor may be placed across a nominally 12V supply rail which under extreme conditions could reach 17V. The difference in power dissipation is nearly double.

3.1.5   Inductance

In some applications, other performance factors must be taken into account. The construction of carbon and metal film resistors is basically a helix cut into a resistive film on a tubular ceramic substrate (Figure 3.1). The dimensions of the helix, together

Figure 3.1 Helical construction of film resistors

with the bulk resistance of the film, determine the actual resistance value. Such a form effectively provides a low-Q inductor in parallel with a low-value capacitor. When the frequency of circuit operation extends into the RF region, the reactance can become a significant part of the total impedance, and non-inductive resistors must be used. Cheapest of these (though increasingly hard to obtain) are carbon composition types, where the resistive element is a homogeneous block of carbon, so that component inductance is primarily that due to the leads. Noise, stability and reliability of these types is poor (which is why they have been superseded for most applications), so for more demanding requirements you need to use specially-wound non-inductive metal film, foil or wire resistors. Alternatively, chip resistors exhibit very low inductance if their handling requirements can be met.

3.1.6   Pulse handling

Another application where conventional helical-cut film resistors are unsuitable is in pulse applications, where a high voltage must be withstood for short durations, so that the average power dissipation is small though the peak power is many times larger. A common application of this type is in thyristor, triac or power transistor snubbers for high-voltage switching. The standard snubber circuit is shown in Figure 3.2.

The RC combination restricts the rate-of-rise of voltage across the device during inductive switch-off, but in so doing the resistor is faced with a momentary fast voltage

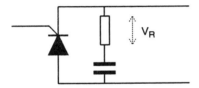 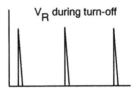

Figure 3.2 The snubber circuit

spike, which in the worst case can approach the power supply voltage. This can cause arcing between adjacent turns of the helix of a film resistor, and swift breakdown of the whole unit. Use of a wirewound high-power component in this application is ill-advised because the high self-inductance of the resistor creates a tuned circuit which, although it has a low Q, can actually *increase* the transient voltage seen by the device. Again, a carbon composition type suffers less from both the breakdown mode and the inductance.

Pulse applications may stress the power rating of the resistor as well as the voltage; the average power dissipated is equal to the peak power times the duty cycle of the pulses, but for pulse duration longer than a millisecond or duty cycles in excess of 10 or 20, the average power permissible has to be derated from the theoretical. Some manufacturers publish curves to allow this derating to be calculated.

3.1.7   Resistor networks

A section on resistors would be incomplete without a mention of the resistor network. There are two main advantages to resistor networks: production efficiency, and value matching/temperature tracking.

Production efficiency

Design for production is a subject in itself, more fully covered in chapter 9. Component handling and insertion costs can be a significant fraction of the total cost of a pcb assembly. A single resistor whose purchase cost is less than 1p may cost as much as 5 – 10p once it has been inserted, depending on how production costs are calculated. If several resistors can be combined in one package then the overall cost of the package plus one insertion can easily be less than the cost of several resistors plus several insertions. To properly evaluate this trade-off you need to have an accurate knowledge of your particular production costs.

Usually resistor networks combine several resistors of one value in a package, dual-in-line or single-in-line, either all separate or with one terminal commoned. An obvious application for the latter type is for digital bus or I/O pull-ups. Linear circuits can benefit as well, especially if they can be designed to use several resistors of one value rather than a mixture of similar values.

Value tracking: thick film vs. thin film

Resistor networks are available using two technologies, the universal thick-film type and the less-widely-available thin film. Thick film networks have no better tolerance and drift specifications than conventional resistors, and so cannot be used in demanding applications. However, because all resistors in a package are constructed by simultaneously screen-printing a resistive ink onto a substrate, manufacturers can guarantee a better tempco tracking between resistors than an absolute tempco for each resistor. A typical performance is 250ppm/°C individually, but a tracking of 50ppm/°C, i.e. all resistors within the package will exhibit the same tempco to within 50ppm/°C. Thin film types can show an order of magnitude better performance.

This feature can be made use of in precision amplifier circuits, and in some instrumentation amplifiers it is essential in order to meet performance requirements. Consider the differential op-amp configuration of Figure 3.3.

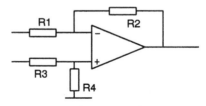

Figure 3.3 Basic differential op-amp configuration

For optimum common-mode rejection the ratios R1/R2 and R3/R4 must be equal; for unity gain all resistor values should be equal. Equality must be maintained over the operating temperature range. While it would be possible to use separate precision resistors of sufficient stability, the absolute value of the resistors is not critical, only their ratios. Since resistor networks can have a much better tempco tracking performance than absolute value, they are highly suited to this type of application. Indeed, some manufacturers offer special networks with different values in the package with guaranteed ratios, especially for such circuits.

Multiples of the package value can be easily obtained by parallel or series connection, and the overall tracking of resistor values is not impaired. If, for example, both half and quarter of the same reference voltage are required, the best circuit for stability and accuracy would have a potential divider in which all resistors were part of the same package (Figure 3.4).

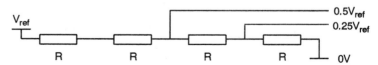

Figure 3.4 Potential divider using equal value resistors

3.2   Potentiometers

Potentiometers are one of the last bastions of electro-mechanical components in the face of the avalanche of digital silicon. They are bulky, unpredictable and unreliable, and can be a problem in fast circuitry; parasitic effects abound, and you can really only use them where the signal frequency times the circuit resistance is less than 10^6 Hz-Ω. They add time and money at the test and calibration stage of the production cycle. Many of their functions can be taken over by microprocessor-controlled digital equivalents. For instance, a classical use for the standard trimpot is to null out that bugbear of analogue amplifiers, the offset voltage. In the days when op-amps had offsets measured in tens of millivolts, this was a very necessary function. But op-amps are now available at reasonable prices with offsets below half a millivolt, and if this is not enough it is often feasible to use a chopper stabilised device whose effective offset is limited only by thermal effects to a few microvolts. Alternatively, the intelligence of a microprocessor can be used to dynamically calibrate out the offset of a cheap op-amp by frequently auto zeroing the entire A-D subsystem. Equally, the gain of a network can be trimmed using a resistive D-A converter rather than a trimpot.

Nevertheless, there will remain applications where a pot is overall the best solution to a given circuit problem: digital implementations have been eschewed for other reasons, or the inherent linearity and lack of distortion of a passive resistor element are essential, or assured non-volatility of a setting is required. For these purposes, you need to consider the different types available.

3.2.1   Trimmer types

Potentiometers can be divided into two classes. Those types which are mounted on the circuit board and are only intended for adjustment on test and calibration, or possibly by maintenance technicians, are known as trimmers and are distinguished by their small size: quarter-inch diameter units are commonplace and 3mm square surface-mount types are now on the market.

Carbon

The cheapest and lowest-performance types have a moulded carbon film track and are of open construction (sometimes known as "skeleton" construction). They are prone to mechanical and environmental degradation and are therefore not suitable for professional applications, but their low cost (approaching 5p in quantity) makes them popular in non-critical areas.

Slightly more expensive (about 20% more) are the enclosed carbon versions which protect the track from direct contamination but are otherwise similar to the skeleton type. Value tolerance for all carbon film trimmers is normally ±20%.

Cermet

The most popular type for commercial and professional purposes is the cermet. Several mounting versions and sizes, all multiple-sourced, are available, with costs ranging from 20p to 80p for single-turn variants. The term "cermet" refers to the resistive element, which is a METal film deposited on a CERamic substrate.

The cermet offers a wide range of resistance, from 10Ω to 2MΩ, with tolerances usually of ±10%, though cheaper parts offer ±20% and tighter tolerances can be ordered. Because it can be obtained in small sizes with low self-capacitance, it is useful at higher frequencies than other types.

Wirewound

The main advantages of wirewound trimmers are their low temperature coefficient, higher power dissipation, lower noise, and tighter resistance tolerance. When used as a variable resistor, their lower contact resistance improves the current-carrying capability through the wiper (cf section 3.2.3). Their resistance stability with time and temperature is slightly better than cermet, but the high-resistance extreme is comparatively low (50kΩ) and the type is not suitable for high frequency. Their resolution is poor and they are also somewhat more expensive and less widely sourced than cermet.

Multi-turn

Both cermet and wirewound trimmers are available in multi-turn configuration. This means that more than 360° mechanical adjustment is needed to cause the wiper to traverse the total resistance element. (Note that a single-turn trimmer normally offers less than 270° adjustment angle.) You will commonly find 4, 10, 12, 15, 20 and 25 turn units. These are used when better adjustability is required than can be offered by a single-turn, but at somewhat greater cost and size, typically twice as much.

3.2.2   Panel types

The other potentiometer classification applies to those which are to be adjusted by the user, and therefore have a spindle which protrudes through the equipment panel. These can be mounted on the panel using an integral bush, or directly to the pcb. The latter tends to put extra strain on the pot's terminals, and requires careful consideration of mechanical tolerances, but is popular because it dispenses with a wiring loom and therefore speeds up production time. The spindle can be insulated or of metal, with or without a locating flat for the knob; insulated types offer a potential safety and EMC advantage, but are also less mechanically rigid.

Carbon, cermet and wirewound

The electrical characteristics of these types are similar to those of single-turn trimmers of the same construction. Because the size of panel pots is necessarily larger to allow for mechanical strength and easy mounting, they can have higher power ratings than trimmers. 0.4W is typical for carbon types, while cermets and wirewounds offer 1 – 5W.

Conductive plastic

When you need a very high-quality track construction, then the conductive plastic type is suitable. These offer very long life and low torque and are also suitable for position transducers. High-accuracy components are very expensive, but general purpose ones can be competitive with good-quality cermet types.

3.2.3   Pot applications

Firstly, remember that the wiper contact is the weakest point. Enormous advances have been made in potentiometer reliability over the years, and the cermet construction has proved itself in many applications. But the wiper is still essentially an electro-mechanical component and as such is the source of virtually all pot problems.

The golden rule is: draw as little dc through the wiper as you possibly can. If you have to draw significant current use a wirewound device. If the pot is used as a variable potential divider to set a circuit voltage, ensure that it is operating into a high impedance. If it is used in a signal path to vary signal amplitude, incorporate a dc blocking capacitor to prevent any dc flow through the wiper. The wiper/element contact has an unpredictable resistance of its own (Figure 3.5) which is affected by oxidation

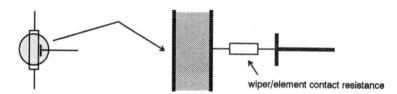

wiper/element contact resistance

Figure 3.5 Potentiometer contact resistance

and electrochemical corrosion, and the effect of this resistance on the circuit (usually manifested as noise) must be minimised. The less current that is drawn through the wiper, the less it will contribute to any noise voltage. Unfortunately, for some types a minimum current (such as 25μA) *should* be drawn through the wiper, in order to "wet" the contact. This should be kept low.

Use as a rheostat

Secondly, if the pot is being used as a variable series resistor (rheostat), connect the wiper to one end of the track. This is a very simple precaution - one short length of pc track. The reason is that under conditions of age, dirt or extreme vibration the wiper can become temporarily (or even permanently) disconnected from the element track. If it is connected as in Figure 3.6(a), the maximum circuit resistance is limited to the end-to-end resistance of the pot. If you connect it as in (b), then the device can become open-circuit. In some circuit configurations this is merely a nuisance, but in others it could be catastrophic.

Figure 3.6 Rheostat connection

Also, remember that pots exhibit an "end resistance" which prevents wiper access to the ends of the resistance element. This restricts the minimum and maximum range that can be obtained: it is impossible to get a potentiometric division ratio (at "minimum volume") of zero.

Adjustability

Do not expect infinite adjustability from your pot. A cermet element is theoretically capable of infinite adjustment, but test and calibration technicians will find a centre-zero reading elusive, and the test and calibration labour costs will multiply, if you try to obtain more resolution from the pot than is realistically achievable. A multi-turn trimmer offers a better solution in this respect (Figure 3.7). Resolution is also compromised by shock- and vibration-induced jumps. If you use a wirewound element, don't even think about high resolution, think of it more like a 100-way resistance switch.

As is emphasized in the section on design for production, one of the major aims for any designer must be to make their design cheap to produce. Any trimming adjustment is a production cost and the time taken to do it must be minimised. The fewer trim adjustments there are, the better the design. But, when you are selecting a trimmer and determining its placement on the board, keep in mind the people who will have to use it and ensure that the adjustment screw is accessible. Place side-adjustment pots on the edge of the board and top-adjustment ones in the middle.

Law accuracy

The two common potentiometer laws are linear and logarithmic. These refer to the law which relates angular displacement to proportion of total track resistance at the slider. For linear pots, the law accuracy is specified as linearity and is closely related to cost; a low-cost panel component will probably not specify linearity at all and it is likely to be no better than 10%. Logarithmic pots are generally only intended for use as audio

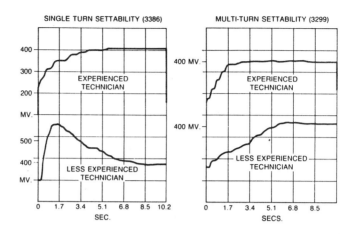

Figure 3.7 Settability of single- and multi-turn trimmers
Source: Bourns

volume controls and the adherence to a log law will be even less accurate. If you need high law accuracy, for instance because you are using the pot as a position transducer, you will need to specify and pay for it. Linearities better than 1% are possible, but a typical high-quality component will offer no better than 5%. Note that the specification tolerance is a different parameter, as it refers only to the tolerance on the end-to-end resistance of the track.

Manufacturing processes

Another factor which works against potentiometers is their dislike of board soldering and cleaning processes. This of course applies to any electro-mechanical component: relays and switches are equally susceptible. When a board is being soldered, it and the components it carries are exposed to extreme temperature shocks; once it has been soldered it carries flux residues which must be removed by washing in water or solvent. Electro-mechanical components can be sealed or open. If they are sealed, there is the danger of a damaged seal allowing ingress of washing fluid which then remains and causes an early failure of the component. If open, the danger is that the washing fluid will directly affect the component's operation. Either case is unsatisfactory, and many manufacturers prefer to add electromechanical components by hand, for reliability reasons, after the rest of the board has been populated, soldered and cleaned – which is clearly an added production cost.

3.3 Capacitors

Like resistors, capacitors can tend to be taken for granted. There is as much a profusion of capacitor types and sub-variants as there is of resistors, and it is often hard to select the optimum part for the application. Table 3.3 shows the characteristics and applications of the more common types.

Capacitors can be subdivided into a number of major types and subheadings within those types. The divisions are best made by dielectric:

Film   Polyester, Polycarbonate, Polypropylene, Polystyrene
Paper

Type	Cap. Range	WV range	Tolerance	Tempco range	Manufacturers	Applications	Unit Cost
Metallised film: Polyester	1nF - 15µF	50 - 1500V	5%, 10%, 20%	Non-linear +/-5% over -55>100°C	Wima, Philips, Siemens,	General purpose coupling & decoupling	5p - 50p
Polycarbonate	100pF - 15µF	63 - 1000V	5%, 10%, 20%	< -1% over -55>100°C	Arcotronics, Evox,	Low tc, timing & filtering	10p - £3
Polypropylene	100pF - 10µF	63 - 2000V	5%, 10%, 20%	+/-2% over -55>100°C	ICW, LCR, Roederstein	High power, high freq.	10p - £1.50
Polystyrene	10pF - 47nF	30V - 630V	1% - 10%	-125ppm/°C	Suflex, Evox, LCR	Close tolerance, low-loss	7p - 50p
Metallised paper	1nF - 0.47µF	250VAC	+/-20%	-	Wima, RIFA	Mains RFI Suppression	20p - 60p
Ceramic: Single layer	10nF - 220nF 1pF - 47nF	12V - 50V 50V - 6KV	-20% +80% 2% > 20/80%	(Barrier layer type) Dependent on dielectric	Beck, Taiyo Yuden, Thomson, Dubilier, Philips, Murata	General purpose & HV	3p - £1.50
Multilayer: COG/NPO	10pF - 27nF	50 - 200V	2%, 5%, 10%	0 +/- 30ppm/°C	AVX, Vitramon, Syfer, Sprague, Philips	Low tc, frequency sensitive & timing	30p - £2 moulded
X7R	1nF - 680nF	50 - 200V	5%, 10%, 20%	Non-linear +/-15% over -55>125°C		General purpose coupling & decoupling	10p - £1.50 dipped
Z5U	1nF - 2.2µF	50V, 100V	20%, -20%+80%	Non-linear: +22%,-56% over +10>85°C			
Electrolytics: Aluminium Oxide	1µF - 4700µF (0.1 - 68000µF)	6.3 - 100V (up to 450V)	+/- 20% (-10% +50%)	-	Philips, Dubilier, NCC, Waycom, Rubycon etc	General purpose reservoir & decoupling	4p - £3 (10p - £10)
Solid Aluminium	0.1µF - 68µF	4 - 40V	+/-20%	-	Philips	High-performance	20p - £2
Tantalum bead	0.1µF - 150µF	6.3 - 35V	+/-20%	-	NEC, Sprague, Dubilier, Waycom, STC	General purpose small size	6p - £1

Notes: 1) This survey does not consider special types 2) Manufacturers quoted are those widely sourced in the UK at time of writing 3) Quoted ranges are for guidance only 4) Costs are typical for medium quantities

Table 3.3 Survey of capacitor types

Ceramic: Single layer: barrier layer, high-K, low-K
                  Multi-layer: COG, X7R, Z5U
Electrolytic: Non-solid and solid aluminium, solid tantalum

This list covers all of the common types likely to be used in general-purpose circuit design. There are certain special or obsolete types – porcelain, trimmer, air dielectric, silver mica – which are not included because their applications are too specialised. The divisions listed above are further subdivided depending on their construction – radial lead, axial lead, disc, chip or whatever – but this does not affect their fundamental circuit characteristics, although it becomes important when pcb layout and production processes are considered.

3.3.1 Metallized film & paper

For a survey of the applications, let us start with the film types. These all have the same general construction, a sandwich of dielectric and conductive films wound into a roll and encapsulated along with their connecting wires, as shown in Figure 3.8. There are

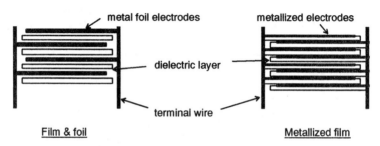

Figure 3.8 Film capacitor construction

two common methods of providing the electrode; one has a separate metal foil wound with the film dielectric, the other has a conductive film metallized onto the dielectric directly. The film and foil construction requires a thicker dielectric film to reduce the risk of pinholes, and therefore is more suitable to lower capacitance values and larger case sizes. Metallized foil has self-healing properties – arcing through a pinhole will vapourise the metallization away from the pinhole area – and can therefore utilize thinner dielectric films, which leads to higher capacitance values and smaller size. The thinnest dielectric in current use is of the order of 1.5μm.

Polyester

Of the film dielectrics listed the most common is polyester. This has the highest dielectric constant and so is capable of the highest capacitance per unit volume. It rivals multilayer ceramic capacitors for volumetric efficiency and in fact can be used in most of the same applications: decoupling, coupling and by-pass, where the stability and loss factor of the capacitor are not too important.

Polyester has a non-linear and comparatively high temperature coefficient. Its dissipation factor tan δ (see Figure 3.9) is also high, of the order of 8×10^{-3} at 1kHz and 20°C, and varies markedly with temperature and operating frequency. These factors make it less useful for critical circuits where a stable, low-loss component is needed.

Polycarbonate

For these cases polycarbonate is best suited. This has a near flat temperature –

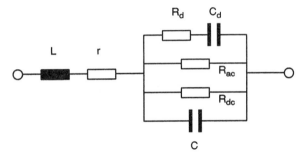

C is the "ideal" capacitance of the device

R_{ac} is the equivalent resistance due to ac dielectric losses, and may vary non-linearly with frequency and temperature

R_{dc} is the leakage or isolation resistance of the dielectric, and may vary with temperature

r is the equivalent series resistance (E.S.R.) due to the electrode, lead and terminating resistances

L is the equivalent series inductance (E.S.L.) due to the electrodes and leads

R_d, C_d are equivalent components to represent dielectric absorption properties

Capacitor Specifications

Temperature coefficient is the change in capacitance C with temperature and may be quoted in parts per million (ppm)/°C or as a percentage change of C over the operating temperature range

tan δ is a measure of the lossiness of the component and is the ratio of the resistive and reactive parts of the impedance, R/X, where $X = \omega C$

Insulation resistance or **time constant** is a measure of the dc leakage and may be quoted as the product $R_{dc} \cdot C$ in seconds or as R_{dc} in MΩ. **Leakage current** is the same characteristic, more usually quoted for electrolytics

Dielectric absorption is a measure of the "voltage memory" characteristic, which occurs because dielectric material does not polarize instantly, but needs time to recover the full charge; it is defined as a percentage change in stored voltage a given time after the sample is taken.

Figure 3.9 The capacitor equivalent circuit

capacitance characteristic at room temperature, with a decrease of about 1% at the operating extremes. It also has a lower tan δ, typically less than 2×10^{-3} at 20°C, 1kHz. Polycarbonate would normally be specified for frequency-sensitive circuits such as filters and timing functions. It is also a good general-purpose dielectric for higher-power use, but in these applications polypropylene comes into its own.

Polypropylene and polystyrene

Polypropylene has a lower dielectric constant than the others and does not metallize so easily, and so gives a larger component for a given CV product. Also it exhibits a fairly constant negative temperature coefficient of –200ppm/°C which restricts its use in frequency-critical circuits, although a defined tempco can be useful for temperature compensation in some instances. Its main advantage is its very low dissipation factor of around 3×10^{-4} at 20°C and 1kHz, almost constant with temperature. This allows it to handle much higher powers at higher frequencies than the other types, so it is suitable for switchmode power supplies, tv line deflection circuits and other high-power pulse applications.

Polypropylene capacitors can also be made to close tolerances and this makes them competitors to polystyrene in many tuned circuit and timing applications. Additionally, both polypropylene and polystyrene have similar temperature coefficients and loss factors (polystyrene tempco -125ppm/°C, tan δ typically 5×10^{-4}). They also both show a better dielectric absorption performance (0.02% to 0.03% – see Figure 3.9) than the other film types, which makes them best suited to sample-and-hold circuits. Both types suffer from a reduced high temperature rating, generally limited to 85°C, though some polystyrene types are restricted to 70°C and some polypropylene types are extended to 100°C.

Metallized Paper

A further capacitor type which is often bracketed with film dielectrics is the metallized paper component. Paper was historically a very widely-used dielectric, particularly in power applications, before the technical developments in plastic films superseded it. The great advantage of plastic films is that they absorb moisture very much less readily than paper, which has to be impregnated to prevent moisture ingress from destroying its dielectric properties. Paper is now mostly reserved for use in special applications, particularly for across-the-line mains interference suppressors. When a capacitor is used directly across the mains, a fault in the dielectric can lead to localised self-heating and eventually the component will catch fire, without ever blowing the protective mains fuse. This has been found to be the cause of many electrical equipment fires and much investigation has been carried out (prompted by insurance claims) into the question of capacitor flammability. Paper has excellent regenerative characteristics under fault conditions; very much less carbon is deposited by a transient dielectric breakdown than is the case for any of the plastic film dielectrics, so that self-heating is minimal and the component does not ignite. Metallized paper is therefore the preferred construction for this application, although polyester, polypropylene and ceramic suppressor types are available.

3.3.2   Multi-layer ceramics

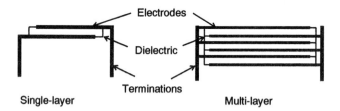

Figure 3.10 Ceramic capacitor construction

Multi-layer ceramics have a superficially similar construction to film capacitors, but instead of being wound, layers of dielectric and electrode material are built up individually and then fired to produce a solid block with terminations at each end (Figure 3.10). For this reason they are often called "monolithic" capacitors. They can be supplied either in chip form or encapsulated with leads; the encapsulant can be a moulded body or a dip coating.

Of the many manufacturers of monolithic multi-layer ceramics, virtually all offer products in three grades of dielectric: COG, X7R and Z5U. COG is also known as NP0, referring to its temperature coefficient. These three classifications are standardized internationally by the IEC (Europe) and EIA (America), so allowing direct comparison of different manufacturers' offerings. The three grades are quite different.

COG

COG is the highest quality of the three but has a lower permittivity, which means that its capacitance range is more restricted. It exhibits a near-zero temperature coefficient, negligible capacitance and dissipation factor change with voltage or frequency, and its tan δ is around 0.001. These features make it the leading contender for high-stability applications, though polycarbonate can be used in some cases and will usually beat it on price.

X7R

X7R is a reasonably stable high-permittivity (Hi-K) dielectric, which allows capacitance values up to 1µF to be achieved within a reasonable package size. It can be used over the same temperature range as COG but it exhibits a non-linear and quite marked change of both capacitance and tan δ over this range. Tan δ at 20°C and 1kHz is 0.025. Capacitance and tan δ also change with applied voltage and frequency by up to 10%, which rules out many applications, really leaving only the general-purpose coupling and decoupling area.

Z5U

In comparison with the previous two, the Z5U dielectric shows a very much worse performance. Its capacitance changes by over 50% with changes in temperature and applied voltage; its rated temperature range is only +10°C to +85°C, though it can be used at lower temperatures. Its tan δ is similar to X7R. Its initial tolerance can be as wide as −20%, +80%. Working voltages are restricted to 100V. Virtually its only redeeming feature is its high permittivity which allows high capacitance values, up to 2.2µF, to be achieved. Its performance limitations mean that its only real application is for IC decoupling; however, this market alone guarantees it sales of millions of units, so it remains widely available.

Ceramic versus film

When comparing the specifications of multilayer ceramics (particularly COG) and metallized film components, there often seems to be little difference between them. Even price and package size come out roughly the same. In fact for most applications they are interchangeable (as long as like is compared with like). There is an interesting geographical factor involved, though. US manufacturers are stronger in ceramics than in film, whereas in Europe the position is reversed. Therefore, US design engineers will tend to design in a ceramic component, while their European counterparts will go for film. So far so good; but the majority of digital ICs sold throughout the world are of US origin, and their application notes are written by US applications engineers. These, quite naturally, specify ceramic capacitors when it comes to recommended practice for IC decoupling. As a result, the suppliers of monolithic ceramics are riding on the back of an enormous hidden subsidy as designers worldwide take the easy route of sticking to recommended parts in their circuit designs.

3.3.3 Single layer ceramics

By contrast, single layer ceramic capacitors are mostly of European or Japanese origin. There is a wide range of thicknesses and types of dielectric material, and so the varieties of capacitor under this heading are correspondingly wide. Barrier layer ceramics use a semiconducting dielectric with a surface layer of oxide, which results in two very thin dielectric layers, effectively connected in series. The thickness of the formed layers determines the capacitance and working voltage, so that for a given disc size C and V are inversely proportional. Breakdown voltage is low, tan δ is high, and capacitance change with temperature, voltage and frequency are all high. Their only practical advantage over Z5U multilayers (see above) is cost.

Low-K and High-K dielectrics

Other single-layer ceramics use low-K (permittivity) or high-K dielectric materials, and are generally distinguished as Type 1 or Type 2 (or Class 1/Class 2).

Type 1 dielectrics have a temperature coefficient ranging from +100 to –1500ppm/°C or greater, depending on the ceramic composition. The tempco is reasonably linear, and the capacitance value is stable against voltage and frequency. Low tan δ at high frequencies (typically 1.5×10^{-3} at 1MHz) allows their use extensively in RF applications. Because of their low permittivity the capacitance range extends from less than 1pF to around 500pF.

Type 2 dielectrics use ferro-electric materials, usually barium titanate, to allow much higher permittivities to be obtained, at the cost of variability in capacitance and tan δ with voltage, frequency, temperature and age. The X7R and Z5U dielectrics used in multilayer components are also available among the very many type 2 materials used for single-layers. Each manufacturer offers their own particular brew of ceramic and if your application requires critical knowledge of C and/or tan δ then inspect the published curves closely. Capacitance ranges from 100pF up to 47nF are common. Working voltages can be extended into the kV region by simply increasing the dielectric thickness, with a corresponding increase in electrode plate area.

3.3.4   Electrolytics

Electrolytic capacitors represent the last major subdivision of types. Within this subdivision the most popular type is the non-solid aluminium electrolytic. These are available from numerous suppliers and are used for many applications. There are two major characteristics common to all electrolytics: they achieve very high capacitance for a given volume, and they are polarised. General-purpose aluminium electrolytics span the capacitance range from 1μF to 4700μF. Large power-supply versions can reach tens of thousands of μF; sub-miniature versions can be had down to 0.1μF, where they compete with film and ceramic types on both size and price.

Construction

The non-solid aluminium electrolytic is constructed from a long strip of treated aluminium foil wound into a cylindrical body and encased (Figure 3.11). The dielectric is aluminium oxide (Al_2O_3) formed by electrochemically oxidising the aluminium. It is contacted on one side by the base metal (the anode) and on the other by an ionic conducting electrolyte contained within a spacer of porous paper. The electrolyte is itself contacted by another electrode of aluminium which forms the cathode contact.

The electrodes are normally etched to increase their effective surface area and so increase the capacitance per unit volume, hence modern electrolytics are sometimes known as "etched aluminium". Because the electrolyte is an ionic conductor the polarity of the applied voltage must not be reversed, or hydrogen will be dissociated at the anode and create a destructive overpressure. The thickness of the Al_2O_3 dielectric

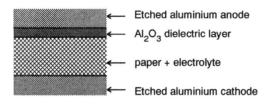

Etched aluminium anode

Al_2O_3 dielectric layer

paper + electrolyte

Etched aluminium cathode

Figure 3.11 Electrolytic capacitor construction

is determined by the applied voltage during the electrochemical forming process, and to avoid further thickening of the dielectric during use (again with destructive results) the operating voltage must be kept below this voltage by a suitable safety factor, which determines the rated voltage of the component. Some manufacturers may specify a "surge" voltage rating, which is effectively the forming voltage without a safety factor.

Solid aluminium electrolytics are a variant in which the cathode is formed from a manganese dioxide semiconductor layer, contacted by an etched aluminium electrode. Although in principle some reverse polarisation is possible with this system, it is not recommended because some ionic reactions due to trapped moisture can still occur.

Leakage

The important electrolytic characteristics tend to be dependent on application, and two broad areas can be discussed: for general purpose coupling and decoupling, and for power supply reservoir purposes. In the first field, the most important factor is usually leakage current, which becomes especially critical in timing circuits where it will determine the maximum achievable time constant. Leakage currents for general-purpose components are usually between $0.01CV$ and $0.03CV$ µA where C and V are the rated capacitance and working voltage. Many manufacturers offer low leakage versions which are usually specified at $0.002CV$ µA. Alternatively, leakage current is a fairly well-defined function of applied voltage and usually drops to around a tenth of its rated value at about 40% of rated voltage, so that a low-leakage characteristic can be obtained by under-running the component. Leakage is also temperature-dependent and can be ten times its rated value at 25°C when run at maximum operating temperature. It is also a function of history (see later under "lifetime"): when voltage is first applied to a new component its leakage is higher.

Ripple current and ESR

For power supply reservoir applications, leakage current is unimportant and instead two other factors must be considered, ripple current (I_R) and equivalent series resistance (ESR). The ripple current is the ac current flowing through the capacitor as the reservoir charges and discharges, usually at 100/120Hz for ac mains supplies or at the switching frequency for switch-mode supplies. It develops a power dissipation across the resistive part of the capacitor impedance (ESR) which results in a temperature rise within the capacitor, and it is this dissipation which limits the capacitor's I_R rating. Published data for all electrolytics include an I_R rating which must be observed. The rating increases to some extent with increasing frequency and reducing temperature. But note that the rating is normally published as an RMS value, and actual ripple waveforms are often far from sinusoidal, so a correction factor must be derived for this difference.

The ESR value (Figure 3.12) is important both because it contributes to the I_R rating and because it limits the effective high-frequency impedance of the capacitor. This

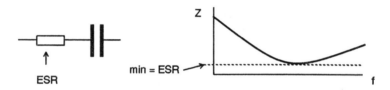

Figure 3.12 Capacitor equivalent series resistance

point has become of increasing importance with the advent of high-frequency switching power supplies where the ripple voltage attainable on the output is limited by the output capacitor's ESR rather than its absolute capacitance value. Some manufacturers now offer special low-ESR versions specifically for these applications. ESR of non-solid electrolytics increases dramatically as the operating temperature is reduced below 0°C, which can be a problem in circuits where actual dissipation is low. The better-quality ranges of electrolytics include this factor in their specification of "impedance ratio", the ratio of ESR at some sub-zero temperature to that at 20°C, which is usually around 3 or 4 but may be much worse. Solid electrolytics do not exhibit this behaviour to the same extent.

Temperature and lifetime

The general capacitor characteristics as discussed for ceramic and film types are generally worse for electrolytics. Capacitance/temperature curves are rarely published but non-solid types can vary non-linearly by around ±20% over the operating temperature range; solid types are better by a factor of two. Tan δ is around 0.1 – 0.2 at 100Hz and 20°C but varies wildly with temperature and frequency. Temperature ranges are normally restricted to –40°C to 85°C, with some types being rated for extended temperature to 105°C or 125°C. Lifetime is an issue with electrolytics, in two respects. Non-solid electrolytics suffer from eventual drying-out of the electrolyte, which is a function of operating temperature and the integrity of the component seal. In general, the life of these types can be doubled for each 10°C drop in operating temperature. Solid electrolytic types do not show this failure mechanism.

The second problem is that of shelf life. Non-solid aluminium electrolytics are one of the few types of electronic component that degrade when not in use. The dielectric Al_2O_3 film can deteriorate, leading to increased leakage, if the component is maintained for long periods without a polarising voltage. The effect is dependent on temperature and shelf life is usually measured in years at 25°C. Capacitors that have suffered this type of degradation can be "re-formed" by applying the forming voltage across them through a current limiting resistor, if this is found to be necessary. At the same time, it is inadvisable to run this type of electrolytic in circuit configurations where it is not normally exposed to a polarising voltage.

3.3.5   Solid tantalum

Solid tantalum electrolytics are generally used when the various performance and reliability limitations of aluminium electrolytics cannot be tolerated. The construction is similar to the solid aluminium type, with a manganese dioxide electrolyte and sintered tantalum powder for the anode. They can be supplied with a temperature range of –55°C to 85°C or up to +125°C, and have a very much greater reliability than aluminium, and so are favoured for military use. Reliability degrades when they are operated in low-impedance circuits and so they are not suitable for power supply reservoir applications, and indeed are not available in high capacitance values, but are very well suited to small-signal coupling, decoupling and timing.

Leakage current is around 0.01 CV μA which is comparable to the better aluminium types, and tan δ is between 0.04 and 0.1, about twice as good as aluminium. Capacitance change with temperature can vary from ±15% to as good as ±3% across the working temperature range. Some proportion of the working voltage can be tolerated in the reverse direction, which relaxes the application constraints. The resin-dipped bead tantalum construction offers usually the best trade-off between price, performance and size in a given application, and tantalum beads are available from a wide range of manufacturers.

3.3.6   Capacitor applications

As with resistors, the actual capacitance that a component can exhibit is only mildly related to its marked value. The art of circuit design lies in knowing which components must be carefully specified and which can have wide tolerances.

Value shifts

The actual capacitance will vary with initial tolerance, temperature, applied voltage, frequency and time.

$$C_{actual} = C_{marked} \times [\pm\text{tolerance}] \cdot [\pm\Delta T \times \text{temp coeff}] \cdot$$
$$[\pm\Delta V \times \text{voltage coeff}] \cdot [\pm\Delta f \times \text{freq coeff}] \cdot [t \times \text{ageing coeff}]$$

Take a nominally 0.1μF Z5U multilayer ceramic capacitor, rated at 50V. It has an initial tolerance of –20%,+80%; a temperature coefficient of +22%, –56% max. over the temperature range +10 to +85°C; a capacitance voltage coefficient that reduces by 35% of its value at 60% of rated voltage; a frequency characteristic that reduces capacitance by 3% of its value at 10kHz and 6% at 100kHz; and an ageing characteristic that reduces its value by 6% after 1000hr. It is run in a circuit with an operating frequency between 10kHz and 100kHz, over its full temperature range, and with an applied voltage that varies from 5V to 30V. The worst case limits of its actual value will be:

a)   0.1μF x 1.8 [max tolerance] x 1.22 [max pos temp coeff] x 1 [max voltage coeff] x 0.97 [max freq coeff]

     = 0.213μF

b)   0.1μF x 0.8 [min tolerance] x 0.44 [max neg temp coeff] x 0.65 [min voltage coeff] x 0.94 [min freq coeff] x 0.94 [ageing]

     = 0.0202μF

In other words, a 10:1 variation. Where is the 0.1μF capacitor now? At the very least, you can see that it is most unlikely to be 0.1μF!

Now repeat the calculation for a 10% polycarbonate component of the same value and 63V rating, which will be larger; and a 20% tantalum bead electrolytic rated at 35V, which will be roughly the same size but polarized. All three types are roughly the same price.

Polycarbonate:

a)   0.1μF x 1.1 [max tolerance] x 1 [max pos temp coeff] x 1 [max voltage coeff]
        x 1 [max freq coeff]

        = **0.11μF**

b)   0.1μF x 0.9 [min tolerance] x 0.994 [max neg temp coeff] x 1 [min voltage
        coeff] x 1 [min freq coeff] x 1 [ageing]

        = **0.089μF**

Tantalum bead:

a)   0.1μF x 1.2 [max tolerance] x 1.05 [max pos temp coeff] x 1 [max voltage
        coeff] x 0.9 [max freq coeff]

        = **0.113μF**

b)   0.1μF x 0.8 [min tolerance] x 0.99 [max neg temp coeff] x 1 [min voltage
        coeff] x 0.5 [min freq coeff] x 0.95 [ageing]

        = **0.038μF**

The polycarbonate performance is dominated by its initial tolerance; the tantalum bead shows a worse performance at the higher frequency. Clearly some 0.1μF types are better than others!

The circuit conditions quoted above are fairly extreme, particularly the wide voltage range and the frequency excursions. But there are some applications where the subtleties of capacitor behaviour have a serious effect on circuit operation. Consider the simple op-amp integrator.

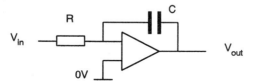

The output voltage follows the law

$$V_{out} = -V_{in} \times t/CR$$

Usually you will assume that C and R are constant and so if V_{in} is also constant the output ramp is linear with time. But if the integrator output swings over a wide range, say the 5 - 30V considered in the last example, C may not be constant but will change as the voltage across it changes. This effect is most marked in accurate timing circuits or in circuits where a linear ramp is used to measure another voltage, such as in some A-D converters. A Z5U ceramic will introduce an enormous non-linearity; even X7R will have a poor showing.

The effect is best combated by choosing the proper dielectric type and/or under-running it, i.e. using a 100V rated capacitor with no more than a 1V ramp. Plastic film would be a better choice, polycarbonate being the preferred type for tempco and frequency stability, and would introduce negligible non-linearity.

NPO/COG ceramic, polycarbonate, polystyrene and polypropylene in roughly that order are most suitable for any circuit which relies on the stability of a capacitor, especially timing, tuning and oscillator circuits, as their capacitance values are least subject to temperature and ageing. Of course, these types are restricted to the lower capacitance ranges (pF or nF – only polycarbonate stretches up to the μF range, and here size is usually a problem) and so are more suitable for higher frequencies. If long, stable time periods are needed it is usually better to divide down a high frequency using a digital divider chain than it is to use large value capacitors, or to expect stability from an electrolytic.

3.3.7   Series capacitors and dc leakage

Capacitor voltage ratings can hide pitfalls. As discussed earlier, it is always better to under-run the working voltage of a capacitor for reasons of reliability. If a particular working voltage is just too high for the wanted capacitor type, it may seem reasonable to simply put two or more capacitors in series and multiply up the overall voltage rating accordingly, always taking into account the reduced total capacitance.

This is certainly possible but more is required than just multiple capacitors. The capacitor equivalent circuit (Figure 3.9) also includes the dc leakage resistance R_{dc} of each capacitor, as shown in Figure 3.13. The dc working voltage impressed across the

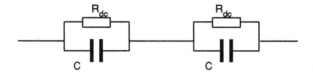

Figure 3.13 DC leakage resistance

terminals is divided between the capacitors not by the ratio of capacitance, but by the ratio of the two values of R_{dc}. These are undefined (except for a minimum) and can vary greatly even between two nominally identical components. Because R_{dc} is usually high, of the order of tens to thousands of megohms, other leakage resistance factors – particularly pc board leakage (see section 2.3) – will also have an effect. The result is that the actual voltage across each capacitor is unpredictable and could be greater than the rated voltage. The problem is at its worst with electrolytics whose leakage current is large and varies with temperature and time.

The situation is to a certain extent self correcting because an over-voltage will in most cases result in increased leakage which will in turn reduce the over-voltage. The major consequence is that the actual capacitor working voltage will be unpredictable and therefore reliability of the combination will suffer. Once one component goes short-circuit the others will be immediately over-stressed and rapid failure of all will follow.

Adding bleed resistors

The solution is simple and consists of placing resistors across each capacitor to swamp out the dc leakage resistance (Figure 3.14). The resistors are sized to be comfortably below the minimum specified leakage resistance so that variations in working voltage are kept below the rated maximum for each capacitor. Naturally this increases the leakage current of the combination but this is often an acceptable price, particularly if the application is for a high-voltage reservoir where some extra drain current is available.

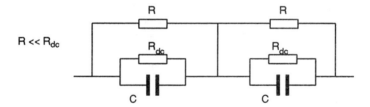

Figure 3.14 Swamping leakage resistance

Indeed, it is often necessary for safety reasons to have a defined "bleed" resistance across a high-voltage reservoir capacitor. If the load resistance is very high, it can take seconds or even minutes for the capacitor voltage to discharge to a safe level after power is removed, with a consequent risk of shock to repair or test technicians. A bleed resistor (or resistors, if the voltage rating of a single unit is inadequate) is a simple way of defining a maximum discharge time to reach a safe voltage.

3.3.8   Dielectric absorption

Another effect mentioned earlier is the phenomenon of dielectric absorption. If a capacitor is charged to a given voltage, discharged by shorting it, and then open-circuited again, its voltage will begin to creep up from zero towards the original voltage. The capacitor exhibits a "voltage memory" because the dielectric molecular dipoles need time to align themselves in an electric field.

This effect is of most concern to designers of sample-and-hold circuits. If a capacitor has been holding a voltage V_A, and then samples another voltage V_B at the other end of its range, when returned to hold mode (effectively open-circuit) its voltage will drift exponentially towards the old voltage V_A (Figure 3.15).

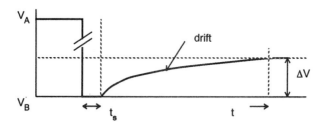

Figure 3.15 Dielectric absorption drift characteristic

The effect can be modelled by an extra circuit in parallel with the main capacitor, $R_d C_d$ in the equivalent circuit (Figure 3.9). $R_d C_d$ has a long time constant and transfers charge slowly to C when the capacitor is open-circuit. The dielectric absorption figure, quoted for $t \gg t_s$, is $\Delta V / V_A - V_B$. The error due to dielectric absorption in a typical circuit will be reduced if the hold time of the old voltage (V_A) is short, or if the measurement is made just after the sample is taken rather than many multiples of t_s later. At the same time, selection of capacitor type plays a part, and of the readily available dielectrics polystyrene and polypropylene are the best, exhibiting dielectric absorption factors of 0.01% – 0.02%.

3.3.9   Self resonance

When capacitors are used at high frequencies another factor comes into play, and this is their self-resonant frequency (s.r.f.). The capacitor equivalent circuit includes the ideal capacitance C, the equivalent series resistance and the equivalent series inductance (ESL). These three components form a low-Q series tuned circuit whose impedance versus frequency curve has the well-known characteristic shape of Figure 3.16, showing a minimum impedance at self-resonance.

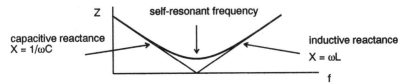

Figure 3.16 Capacitor self-resonance

The term "high frequency" here is relative. All capacitors exhibit the same basic curve, but the minimum for say a 47μF tantalum electrolytic could be at 500kHz and the null could be very flat, or for a 100pF COG chip ceramic the null could be at 100MHz and very well-defined.

The ESL is determined by the lead length and the body size. (Lead length includes the lengths of connecting pc track to the adjoining circuit nodes.) Therefore small, leadless chip capacitors show the lowest inductance and highest self-resonant frequency while large, leaded capacitors have high inductance and low s.r.f. Many manufacturers publish impedance/frequency curves if they expect their components to be used at high frequencies. A typical comparative set of curves, comparing small polyester films, tantalum electrolytics and multilayer ceramics, is shown in Figure 3.17.

Consequences of self-resonance

A capacitor used above its self-resonant frequency is effectively a low-Q inductor and so rf circuits using inappropriate components can show somewhat unpredictable behaviour. The problem also appears more frequently as the speed of digital circuits increases, and the clock frequencies approach or exceed the s.r.f. of the capacitors that are used for supply rail decoupling (*cf* section 6.1.4). Clearly a tantalum electrolytic with a s.r.f. of 1MHz is not much use for decoupling clock transients of 10-20MHz, but you can parallel large-value electrolytics with small-value, e.g. 10nF, ceramic or film capacitors whose s.r.f. is in the 10 - 100MHz region. The combination is then effective over a much wider range of frequencies, and the technique is common practice in wideband amplifier circuits.

3.4   Inductors

Hardly surprisingly, the inductor equivalent circuit (Figure 3.18) is very similar to the capacitor's. All practical components contain the three passive circuit elements, R, C and L, in their makeup. Both capacitors and inductors are fundamentally energy storage

Figure 3.17 (overleaf)   Impedance-frequency characteristics for plastic, ceramic and solid
                                                tantalum capacitors
                                                Source: Waycom

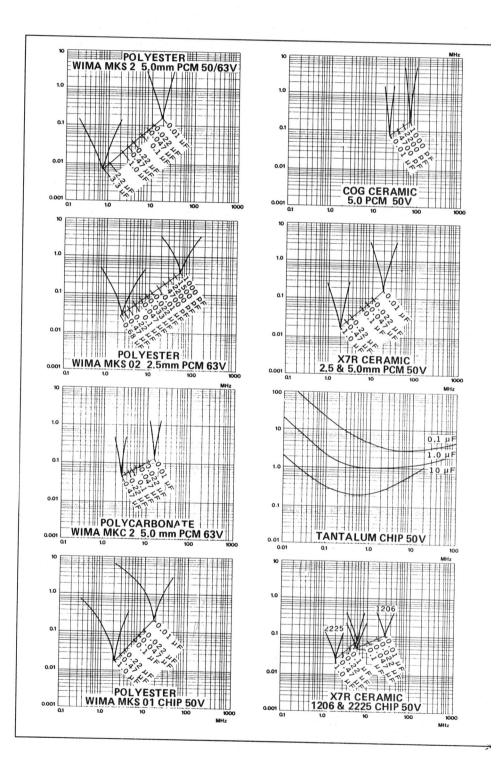

devices, but the practical inductor departs further from the ideal than does the practical capacitor, and so is less universally used. Also, off-the-shelf inductors are less common than capacitors, and the prospects for multiple sourcing are slight.

For this reason a tabular survey like those already given for resistors and capacitors is not really feasible. Inductors are generally made by winding wire around a magnetically permeable material and it is the performance of the material, along with the dc resistance of the wire, which determines the performance of the finished component.

3.4.1   Permeability

Permeability is the measure of a material's ability to concentrate lines of magnetic flux. Air (and other non-magnetic materials) has a relative permeability, μ_r, of 1. Inductors wound on a non-magnetic core are inherently low-loss, the only losses being due to the wire resistance. Unfortunately they are also inherently low-inductance, which generally limits their use to the hf and vhf region. Achieving an inductance greater than $100\mu H$ without a magnetic core requires a large number of turns, so that either the component becomes unmanageably large or the wire becomes unmanageably thin. At low frequencies the winding resistance approaches the inductive reactance, making the coil practically useless. Large air-cored coils do have applications where a stable, low-loss inductance is needed in the presence of high dc currents.

Most inductors for low- and mid-frequency use call for higher values, and must use permeable cores to achieve this. Relative permeabilities of several thousand are possible, allowing small inductors of several Henries to be wound fairly easily. However, there are disadvantages, as you might expect:

- high-μ materials introduce losses of their own, reducing achievable Q factors;

- the materials reduce in permeability as the magnetic field increases. This is known as "saturation", and means that the inductance drops off at high power levels or bias currents;

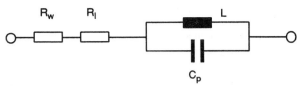

L is the "ideal" inductance of the device
R_w is the series resistance due to the winding wire and terminations, increasing with temperature
R_l is an equivalent series resistance due to the magnetic core losses, and is variable with frequency, temperature and current
C_p is the self-capacitance of the winding, determined by the method of construction of the component
(R_w and R_l may be lumped together into an equivalent series resistance R_{eq}, in which case the overall lossiness of the component is quoted as its Q, which is given by $\omega L/R_{eq}$). The reciprocal of Q is tan δ, which is analogous to the capacitor tan δ.

Figure 3.18 The inductor equivalent circuit

- a slight change in the material's molecular structure on magnetization results in "hysteresis", which shows itself as a remanent magnetic flux when the magnetic field reduces to zero, depending on the level of previous magnetization;

- the absolute magnetic and physical properties of the core are hard to control closely, so that the tolerance on the inductance value of the finished component is wide, though this can be controlled by an air gap in the magnetic circuit;

- the permeability and loss vary with temperature; above the "Curie point" the magnetic properties vanish almost completely.

Figure 3.19 shows a typical curve of magnetic flux versus applied magnetic field (the "B-H" curve) for a ferrite material, illustrating the effects of saturation and hysteresis. You will use such published curves in inductor design, primarily to determine the power-handling capability of the component.

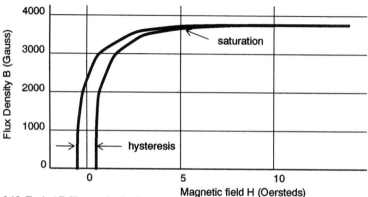

Figure 3.19 Typical B-H curve for ferrite

Ferrites

The properties of permeable cores are reminiscent of the dielectric characteristics - variability with voltage, temperature and frequency - of the hi-K ceramics that we looked at earlier, and indeed there is much in common. The most common core material, ferrite, is in fact a type of ceramic, and is produced in at least as many varieties as are the capacitor ceramic materials. Ferrites are a metal-oxide ceramic made of a mixture of Fe_2O_3 and either manganese-zinc or nickel-zinc oxides pressed or extruded into a range of core shapes. Every manufacturer offers a wide variety of shapes and will usually also offer a custom service for large volumes, but for most uses a selection from a relatively small range of standard types is adequate, and offers the benefit of sourcing from different suppliers. Two of the most popular core types are the RM series to IEC431 and the E, EP and EC series for medium-power high frequency transformers.

Manganese-zinc ferrites have a high permeability but also their losses increase rapidly with frequency, making them more suitable to low-frequency applications. Nickel-zinc ferrites are of lower permeability, but their lower high-frequency losses make them useable up to about 200MHz. Their resistivity is higher by several orders of magnitude, and their Curie point temperature is higher. The ratio of manganese to zinc or nickel to zinc in either case determines the grade of material.

Iron powder

The other core material in wide use is iron powder. Such cores are pressed from a very fine carbonyl iron powder mixed with a bonding material. Eddy current losses in the core are minimised by creating an insulating layer on the surface of each particle before pressing, but this introduces minute gaps in the magnetic circuit, which restricts the permeability of the material to a maximum of around 30. The same effect makes iron powder cores very hard to saturate. The main uses for iron powder are for high frequency tuned circuit cores, and suppressor chokes where low saturability is more important than high inductance.

3.4.2   Self-capacitance

You should never wind directly onto the core. Apart from mechanical instability, the very high dielectric constant of the ferrite will increase the self-capacitance C_p of the winding several times if the two are in close contact. The bobbin serves to keep the winding well spaced from the core and minimizes self-capacitance. The way the winding is built up also influences self-capacitance; a single layer has the lowest capacitance, but if multiple layers are necessary then two possibilities exist to reduce it:

- "scramble" or "wave" wind rather than build up in discrete layers. This can reduce C_p by around 20% over a layer winding, but uses more winding space;

- use a multi-section former for a single winding. Again this uses more space, but for example a two-section former can reduce C_p by a factor of 3.

3.4.3   Inductor applications

There are three major applications for inductors: as frequency determining components in tuned (resonant) circuits, as energy storage components, usually in power supplies, and as filter components in suppression circuits. Each application emphasizes different inductor characteristics and calls for a different approach to inductor design.

Tuned circuits

Signal tuned circuits demand predictable inductance values and high Q (low losses). They do not normally see high bias currents, so core saturation and hysteresis loss is not a problem. For low frequencies, ferrite pot cores are the most popular, but for rf use (above 1MHz or so) other types of ferrite core, or iron dust cores, are better. For the best stability and initial tolerance, a lower permeability material is preferable.

As well as the intrinsic stability of the material, for these applications it is important to consider mechanical stability. Any movement or distortion of the core, or movement of the winding relative to the core, will affect the magnetic path and hence the inductance. Also, any mechanical, magnetic or thermal shock to the core causes an immediate change in permeability followed by a long, slow relaxation towards the original value. This is known as "disaccommodation". These effects mean that the core characteristics have to be very carefully considered when a stable inductance is required in a high-shock or high-vibration environment. It is common for the winding, bobbin and core to be encapsulated in varnish to enhance mechanical stability. Hard encapsulating compound should not be used as the high shrinkage could mechanically damage the brittle core.

Power circuits

Energy storage chokes and power transformers, as used for example in switching power supplies, have a quite different set of important parameters. In these, inductance stability is not required but high volumetric efficiency is. Energy stored in the choke is given by $L \cdot I^2$ and so a material which shows a high saturation flux density, allowing a higher magnetizing current, is to be preferred. At higher operating frequencies hysteresis becomes the dominant loss mechanism, and limits the power handling capacity of the core. A small gap in the magnetic circuit, usually obtained by grinding away a part of the core, allows higher saturation at the expense of lower effective permeability. These considerations point to the use of gapped manganese-zinc ferrites or iron dust cores in which the air gap is inherent in the material.

Suppression

In contrast to the previous applications in which low core losses were required for high Q or high power handling, suppression chokes work best if they have high losses. A suppression circuit has to reflect or absorb high-frequency interference energy and prevent it from being propagated beyond the suppressor. The more energy is absorbed within the choke the better will be its circuit performance. Clearly, high-loss ferrites are the best type for these applications; all ferrites when used well above their intended frequency range exhibit high losses, but materials specifically designed and characterised for this purpose are now being marketed. The ferrite bead (Figure 3.20) is an extreme example, in which a straight piece of wire is transformed into a high-frequency choke merely by stringing a bead onto it. The losses induced in the ferrite at high frequencies give the assembly a complex impedance (resistance + reactance) of several tens of ohms.

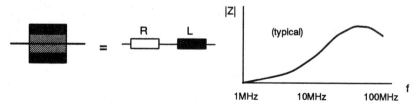

Figure 3.20 The ferrite bead

3.4.4   The danger of inductive transients

One of the fundamental circuit laws pertaining to inductors is the relationship

$$V = -L \cdot di/dt$$

This says, in effect, that the voltage across an inductor is proportional to the rate of change of current through it, and it is the basis for an enormous number of unreliable circuit designs.

Consider a simple circuit: an inductor in series with a resistor and a switch, connected to a dc voltage source, as in Figure 3.21.

When the switch is closed, the current through the inductor will build up according to the above equation until it is limited by R to the steady-state value of V/R. So far so good. But what happens when the switch opens?

The current through the inductor is cut off instantaneously with nowhere to go. But

Figure 3.21 Series inductor-switch circuit

this means that dI/dt is infinite. So, according to the equation, should be the voltage across it. And this is indeed what happens; at the instant of switch-off, a large voltage transient is induced across the inductor. In practice, its amplitude is not infinite but is determined by the Q and the self-capacitance of the inductor, which forms the only path (apart from leakage) for diversion of the stored current. Or, if the self-capacitance is small and the transient is large enough, its amplitude is limited by breakdown across the switch. This is the source of the unreliability of any circuit design which uses this configuration, of which there are many. (It is also the principle of operation of the car ignition circuit, of which there are millions.) As an example, consider the relay coil.

Relay coils

A typical relay drive circuit has the coil driven by a transistor switch (Figure 3.22). The

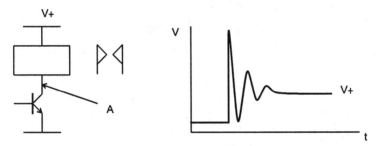

Figure 3.22 Relay driver and turn-off waveform

waveform seen at A on switch-off is a damped sinusoid that has a high initial value but dies away rapidly. The equivalent circuit of the relay coil is an RLC tuned circuit where the R, L and C values are the winding resistance, inductance and self-capacitance respectively. The period of the damped oscillation is determined by the L-C resonant frequency, and the amplitude and decay time constant of the waveform are determined by the Q of the circuit, i.e. by R , L and C. The resistance of a relay coil is always specified by its manufacturer; the inductance is sometimes specified, the self-capacitance never. So the only way to determine the peak value of the transient is to measure it in circuit.

Transient amplitudes of several times the supply voltage are common. It is quite possible to get transients of hundreds of volts from a 12-volt supply, as automotive electronics designers know only too well (*cf* Figure 8.2). If the transistor switch has a collector-emitter breakdown voltage less than the transient peak, the transistor will suffer avalanche breakdown and the transient will be limited. Transistors can withstand repeated low-energy avalanche breakdowns before failure and so, if the circuit is not fully investigated, it will appear to work quite satisfactorily on the bench; it will only be after some time in the field that a high incidence of transistor failures will be noticed.

By that time, the product's reputation for unreliability has been established.

This problem is not limited to transistor drive circuits. There are many applications where a small switching contact is driving a larger switching coil, for example a reed relay driving a larger relay, or a large relay driving a contactor. The same inductive spike phenomenon will lead to spark erosion of the low-power switching contacts, and early failure, if it is not prevented.

Transient protection

Unfortunately, all protection methods involve extra components. Each method seeks to divert away the energy-storage current from the inductor without creating a large transient voltage. The diode method of Figure 3.23(a) is the simplest and often the best. It clamps the positive-going spike to the supply rail without affecting the switching action, and for all practical purposes limits the positive voltage seen by the switch to that of the supply. The diode need only be sized to withstand the surge of flyback current limited by the coil resistance, and its voltage rating need be no more than the supply.

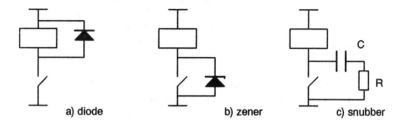

Figure 3.23 Inductive transient protection

Protection against negative transients

The single diode clamp does not protect against a negative-going transient that drives the switch voltage below the 0V rail. This can be prevented by including another diode in series with the switch, or by the zener circuit shown in Figure 3.23(b). The zener clamps both negative and positive transients, but its positive clamping action is somewhat less effective than the single diode. This is because its specified breakdown voltage must ensure that worst-case tolerances on supply voltage and zener voltage do not result in continuous conduction, and so the actual clamping voltage is inevitably higher than optimum (cf section 4.1.8). This is not usually a problem unless the breakdown voltage of the switch is already close to the supply voltage, which is bad practice anyway. The zener has the additional advantage of protecting the switch against transients on the supply rail.

AC circuits

Both the diode and zener methods are only applicable when the supply is a polarised dc voltage. An ac coil requires a different protection method, and for this the snubber circuit Figure 3.23(c) is used. This essentially places an RC network in the path of the inductive current – the network can equally well be across the switch or the coil, provided the supply impedance is low – so that the current is absorbed by the action of charging C. The C is effectively in parallel with the self-capacitance of the coil which it swamps. The resistor limits the switch current when C is discharged at turn-on.

In this circuit, C should be sized carefully to ensure it is no greater than required to reduce the transient to a manageable level, since it slows down the response of the switch and also allows some current into the load when the switch is open. Similarly R should be kept as high as possible consistent with the snubbing action since power is lost both in it and in the switch, as C is discharged. The snubber is a popular network both for ac inductive clamping and in many other circuits as a dV/dt limiter. Calculation of snubber values is outlined in section 4.2.6.

3.5   Crystals

The quartz crystal has been widely used as a frequency-determining component for many years. It is small, robust, accurate and stable. Also, like other components, it has its vices. This section will look briefly at crystal theory before commenting on some application pitfalls.

Quartz (silica, SiO_2) exhibits a piezoelectric effect whereby mechanical stress generates a directionally related electric field, and conversely an applied electric field causes a directionally related force across the crystal. An alternating voltage applied to the crystal will cause it to vibrate, and if its frequency is close to the mechanical resonance the generated electric field will be amplified and can be used to stabilise the applied frequency.

Angle of cut

The crystals used in electronic circuits are in the form of plates or elements cut from a synthetic crystal. The resonant properties vary depending on the angle of cut referred to the base crystal's major axis. X- and Y-cut units, where the direction of cut is perpendicular to the major axis, show subsidiary responses which can reduce the Q-factor of the element and impose a fairly low upper limit on the achievable frequency range. Also, the temperature coefficient of these cuts is large.

Happily, a particular angle of cut of 35° 21′ from the major axis known as the AT-cut, shows very small coupling between the principal and other modes of vibration, therefore lacks subsidiary resonances and is capable of very high-frequency operation. Its resonant frequency is governed directly by a fraction of the element thickness, and its temperature coefficient follows a cubic-plus-linear law whose actual slope varies according to the deviation of angle of cut (see Figure 3.27). The AT-cut crystal is the most widely available crystal unit for general purpose use. Other cuts are available for specialised applications.

3.5.1   Resonance

The crystal equivalent circuit is a series LCR tuned circuit together with a parallel

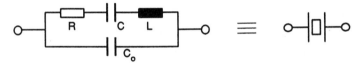

Figure 3.24 Crystal equivalent circuit

capacitance (Figure 3.24). C, L and R are functions of the mechanical resonance properties of the crystal element and C_o is the static capacitance due to the electrodes and terminations. C is very low (of the order of femto-farads) and L is very high (of the order of Henries), while R is generally tens or hundreds of ohms for high-frequency units, and the Q of the resulting combination is very high (30,000 – 100,000). Because of this, the phase angle changes very rapidly with changes in frequency near resonance. So as an oscillator feedback component, the crystal will correct amplifier phase deviations with only a slight frequency shift.

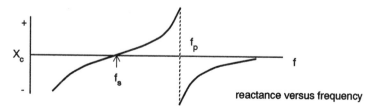

Figure 3.25 Series and parallel resonance

C_o is several hundred times larger than C, and is also increased by external circuit capacitance. The crystal shows two resonant modes, series and parallel, as Figure 3.25 indicates. Their resonant frequencies are very close together and are given by

$$f_s = 1/2\pi \cdot (LC)^{0.5}$$

and

$$f_p = 1/2\pi \cdot (LC_x)^{0.5}$$
where C_x is the series combination of C and C_p, and $C_p = C_o$ + external capacitance

The crystal can be operated in either mode. In series mode, the element is operated at a low impedance which is equivalent to R; it can be "pulled" upwards in frequency slightly from f_s by inserting external series capacitance. In parallel mode the element operates at high impedance and can be pulled downwards in frequency from f_p by adding external parallel capacitance. The frequency shift from f_s in either case is the same for a given external capacitance. Clearly a given resonant frequency is only obtainable if the external capacitance is known, and in fact all crystals are supplied for a quoted "load" capacitance. The unit will only operate at the marked frequency (within the given tolerance) if the actual circuit capacitance is as specified. Conversely, the frequency can be trimmed if absolute accuracy is needed by using a variable load capacitance.

3.5.2   Oscillator circuits

There are two common circuits for digital clock oscillators (Figure 3.26), one operating in each mode. The parallel circuit is only suitable for high-impedance (CMOS) devices while the series circuit can be used for high- or low-impedance devices. The parallel circuit can be run at very low power levels (down to 1μA) but is slow to start. It is commonly used by on-chip microprocessor clock oscillators and other CMOS oscillator/divider ICs, such as real-time clocks.

R_f biases the inverter to linear operation and should be low enough for input bias current to have negligible effect but high enough not to load the crystal. Generally 10–15MΩ is reasonable. The crystal appears primarily inductive and provides 180° phase shift in the feedback loop. C1 and C2 in series together with circuit strays (amplifier

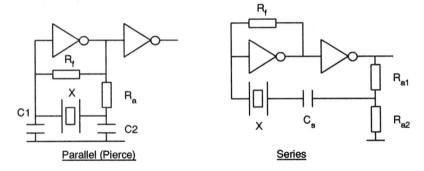

Figure 3.26 Typical crystal oscillators

input and pc track capacitance, amounting to at most 10pF with good layout) form the crystal load capacitance. The ratio C2:C1 should generally be of the order of 3:1, C2 being variable if frequency trimming is desired.

Drive level resistance

R_a is an important component and should not be omitted without proper consideration. It sets the drive level to the crystal. Too high a drive will lead to frequency instability and possible damage to the element. Too low a level will make the oscillator slow to start, perhaps impossible to start with low-activity units, and susceptible to interference. Typical AT-cut crystals have a maximum drive level of 0.5 to 1mW. Some circuits (for example low-power on-chip CMOS oscillators) have a high enough output impedance to make R_a unnecessary but this is not normally the case with discrete-gate oscillators. For watch-crystal units (32.768kHz) R_a should be tens or hundreds of kΩ.

Series circuit

Some applications may be embarrassed by the slow starting time (possibly up to 1 second) of the parallel oscillator circuit. Crystals have a very high Q and if the drive level is low, for frequency stability or to conserve current, the time taken to reach working level is appreciable. This may be unacceptable in microprocessor clock circuits where the clock is expected to be present immediately on power-up. For these purposes the series oscillator, in which the crystal is operated at a low impedance with minimal phase shift across it, is preferable. Its main disadvantage is its higher supply current. The same strictures on drive level apply. Note that the effective series resistance of the element, which is equivalent to its motional resistance R, can vary widely from unit to unit. A spread in this parameter of two- or three-to-one is not uncommon, so it is wise to design the circuit for assured start-up with a three times higher R than quoted.

Layout

Circuit board layout is important, particularly for the parallel mode. Extra capacitance across the crystal should be minimised, as this will increase loop gain and short-term stability. So should coupling between the oscillator circuit and other circuits, especially logic switching circuits, as this decreases the likelihood of spurious oscillation. Ground traces around the crystal to buffer other tracks are advisable; on no account route logic signals near or through the oscillator circuit as they will couple into the high-impedance nodes and cause frequency instability or jitter.

3.5.3 Temperature

Lastly, beware of temperature coefficients. The temperature law of the AT-cut is cubic (Figure 3.27) and can be fairly flat at room temperature, but rapidly worsens as the temperature limits are neared. A crystal will oscillate outside its rated temperature (usually) but the frequency stability will be impaired.

Tuning-fork crystals (the ubiquitous 32.768kHz type, universally used for real-time clocks) show a parabolic curve, of around -0.04ppm/$^\circ$C^2. The turnover temperature is around 25°C which means that for digital watch applications, where the wrist temperature remains around this value, it is ideal and very stable. Transfer this type of crystal to an industrial real-time clock (for example) and its timekeeping at the extremes of the range is hopeless: at +85°C, and at -35°C, it is 144ppm low which represents a loss of 12 seconds per day. Be warned: use an AT-cut!

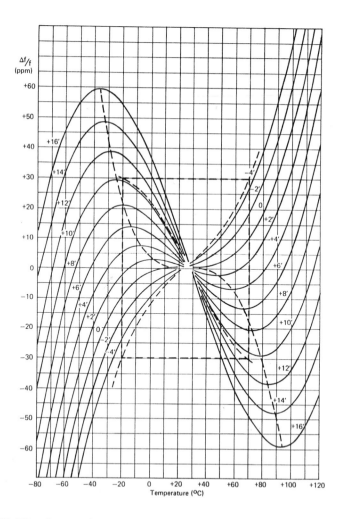

Figure 3.27 AT-cut frequency/temperature curves
Source: ECM

Chapter 4

Active Components

In this chapter we shall concentrate on discrete semiconductor components that are in wide use throughout electronics design. Although there is a continuing trend towards "gathering up" discrete semiconductors into application-specific ICs (ASICs), and a parallel trend towards replacing as many analogue functions as possible by digital signal processing, discrete analogue circuits are still needed when these solutions are impossible or uneconomical. It is as well to be familiar with the characteristics of practical components as even when integrated they show the same fundamental properties.

This chapter covers the common two- and three-terminal devices: diodes, thyristors, triacs, transistors and FETs.

4.1   Diodes

The diode is a two-terminal device whose function is to pass current in one direction but not in the other. A conventional diode is formed from the junction of p-type and n-type silicon. The ideal device has a "brick-wall" V-I characteristic: the practical silicon diode has an exponential characteristic which approximates to the brick wall, if viewed on a large enough scale (Figure 4.1).

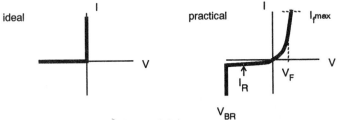

Figure 4.1 Ideal and practical diode characteristics

4.1.1   Forward bias

The first thing to notice is that the forward voltage V_F is not constant, nor is it zero. It has two determinants, forward current I_F and temperature T. They are related by the equation

$$I_F = I_s [\exp (V_F \cdot q/kT) - 1]$$

known as the "diode equation" or the "Ebers-Moll equation", arguably the most fundamental mathematical expression in the whole of semiconductor electronics. The parameters q and k are the electron charge $1.6 \cdot 10^{-19}$ coulombs and Boltzmann's constant $1.38 \cdot 10^{-23}$ joules per degree Kelvin respectively. T is the absolute temperature. The expression kT/q evaluates to 0.025V at 20°C and is the source of many of the properties of the silicon p-n junction. I_s is the saturation current and depends on the device, and on temperature.

A frequent rule of thumb allows 0.6V for the V_F drop of any silicon diode. This is because over the most common range of I_F, V_F remains fairly close to this value. But if I_F is in the microamp or nanoamp region, V_F is close to zero and the slope resistance V_F/I_F is high. The diode behaves more like a nonlinear resistor than a rectifying device. At room temperatures the slope resistance can be taken to be $0.025/I_F$ ohms. If I_F is nearing its maximum limit, which for low-level signal diodes is in the hundreds of milliamps region, V_F can approach 1V and the device begins to dissipate significant power.

Forward current

The maximum forward current, I_{Fmax}, is limited by power dissipation, $I_F \cdot V_F$. This leads to a rise in junction temperature, which must not exceed some maximum value, usually between 125°C and 200°C. Diodes are rated for continuous use but it is possible to exceed the rating for pulse applications, when the average power dissipation depends on the duty cycle ($P_{avg} = D \cdot P_{pk}$). Rectifier diodes are also characterised for "surge" current, which can exceed the average current by 30 – 70 times. The specification is linked to a typical surge duration. Frequently, for US-manufactured rectifiers, this is quoted at 8.33ms, which is one half-cycle at 60Hz, the US mains frequency. If no current-time curves are shown the value can be extrapolated to a limited degree for other time values by taking a constant I^2t product. This specification is important when considering power supply switch-on surges, when a reservoir capacitor is being charged from zero. Remember that V_F carries on rising with I_F as I_{Fmax} is exceeded.

Temperature dependence of forward voltage

The other determinant of forward voltage is temperature. I_s in the diode equation has an exponential temperature dependence which dominates the voltage temperature

Confusion is often caused by the variety of diode symbols in common use. This list shows the accepted symbols (from BS3939/IEC617) for the most frequent types.

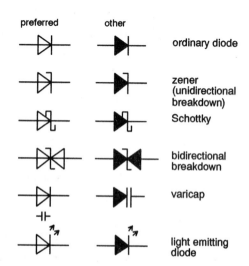

preferred   other

ordinary diode

zener (unidirectional breakdown)

Schottky

bidirectional breakdown

varicap

light emitting diode

Figure 4.2 Diode symbols

coefficient of the device, which for silicon is around $-2\text{mV/}^\circ\text{C}$ for a constant current. This characteristic has numerous desirable applications, and some undesirable effects. It means, for instance, that the $V_F : I_F$ relationship is not a straightforward exponential, because current flowing through the device heats it up, resulting in a complex interdependence between temperature and current. For this reason V_F/I_F curves are normally specified as "instantaneous" and are measured under pulse conditions, which can lead to confusion if these curves are taken to apply to steady-state operation.

The voltage temperature coefficient makes it impossible to assume a stable value for V_F even if the diode is run at a constant current. This has implications whenever a diode is used in a linear circuit.

As a simple example, you may want to use a diode to give a rectifying (unipolar) characteristic to a potential divider. The potential divider relationship is immediately

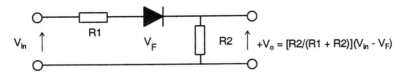

complicated by the addition of V_F. Over the commercial temperature range of 0 to 70°C it will vary by about 150mV. If R1 and R2 are 10K and V_{in} is +5V, V_F is taken to be 0.45 – 0.6V, then V_o will vary from 2.275V to 2.2V over the temperature range – it is _not_ half of V_{in}!

On the positive side, the silicon diode junction does form a cheap and fairly reproducible, if somewhat inaccurate, temperature sensor. Also, two junctions in close proximity can be expected to track changes in V_F repeatably, which allows for fairly simple temperature compensation when necessary. This characteristic is common to all silicon p-n junctions so that, for example, you can use a pair of silicon diodes to compensate the dc conditions of a single-transistor gain stage, as in Figure 4.3.

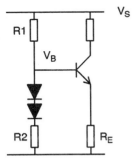

As long as the biasing resistors R1 and R2 are equal, the forward voltage of the diodes ($2V_F$) compensates for the base-emitter voltage of the transistor V_{BE} such that the emitter current is set solely by R_E:

$$V_B \sim [R2/(R1 + R2)] \cdot [V_S - 2V_F] + 2V_F$$

$$I_E = (V_B - V_{BE}) \cdot R_E$$

$$= ([R2/(R1 + R2)] \cdot V_S) \cdot R_E$$
$$\text{if } V_{BE} \sim V_F \text{ and R1 = R2}$$

Figure 4.3 Temperature compensation using biasing diodes

Note that this circuit requires two diodes and the bias resistors to be equal, since the combined forward voltage is divided by the resistor ratio. The compensation is not accurate, because the diode and transistor junctions are not at identical temperatures, and they do not generally carry the same current. If R1 >> R2 then it is possible to get away with one diode and accept rough-and-ready temperature compensation, which may be adequate for your application. Alternatively, use dual transistors to ensure identical junction temperatures, with a more complex circuit arrangement to achieve very accurate compensation.

4.1.2   Reverse bias

So far we have only considered the forward characteristic, that is for positive applied voltage. An ideal diode would block all current flow in the reverse direction. A practical diode doesn't. There are two main reverse characteristics, reverse leakage current I_R and reverse breakdown voltage V_{BR}. The diode equation holds good in the reverse direction until V_{BR} is approached; in the low-voltage region I_R is almost equal to I_S.

Breakdown

V_{BR} is that voltage at which the reverse-biased junction can no longer withstand the applied electric field. At this point, avalanche breakdown occurs and a current limited mainly by the external source impedance will flow. If the device maximum power dissipation is exceeded the junction will be destroyed. Diodes operated conventionally, as opposed to Zener diodes to which we will return shortly, are always run at reverse voltages lower than V_{BR}.

A common over-voltage excursion is the inductive turn-off transient (see section 3.4.4) where a diode is used, intentionally or not, to block the transient. It can be difficult to predict the maximum voltage of the transient and, since the energy dissipated by a breakdown may be much less than needed to destroy the diode, such breakdowns may go unnoticed during the evaluation of the design. Diodes are available which are characterised for the amount of avalanche breakdown energy they can withstand, and should be used if a circuit is expected to deliver predictable transients above the normal breakdown voltage.

4.1.3   Leakage

Since reverse leakage current is of fundamental importance to circuit operation, all diode data sheets quote a specification for maximum leakage. Unfortunately, this hides as much as it reveals.

Leakage current I_R is relatively constant with voltage until V_{BR} is approached, at which point it starts to increase rapidly. It is not, however, constant with temperature, but roughly doubles for every 10°C rise of junction temperature. This characteristic, like the forward voltage temperature coefficient, is common to all reverse-biased pn junctions and we will meet it again later. Most diode leakage currents are specified both at 25°C and at a higher temperature, and the 25°C figure is highly misleading if you apply it over a typical temperature range. A leakage of 100nA at 25°C translates to 2.2µA at 70°C, for instance. This is a common factor in the poor high-temperature performance of high impedance circuits, or those which employ very low current levels.

Leakage variability

To make matters more complicated, leakage is susceptible to process variations. It can vary by up to an order of magnitude from batch to batch under otherwise identical conditions. Therefore, manufacturers will put an artificially high maximum value of leakage in their specifications compared to the actual performance of the majority of delivered units, in order to have room for manoeuvre when a given batch shows a high value.

The consequence of this is that if your design is sensitive to leakage current then the prototype may work well while a production model does not. The probability is high that a device selected at random for the prototype will have a low leakage, whereas some production devices will come from high-leakage batches. If the design proceeded on the basis only of satisfactory measurements on the prototype then the seeds have been sown for production difficulties. To avoid them, always work on worst-case calculations even though these are not borne out by bench tests.

By way of illustration, a set of measured leakage characteristics are shown in Figure 4.4. Several samples each of three types of diode were submitted to a temperature sweep from 0 to 100°C while their leakage currents were monitored. Two different manufacturers' versions of 1N4148, and one version of 1N4004, were tested. The curves show clearly the logarithmic relationship; all the samples had less than 10nA leakage at 25°C. It is also clear that within a batch the variations are quite small, but between two manufacturers of a nominally identical device they are much larger. It is also interesting to see that the 1N4004 rectifier diodes, run at much less than their breakdown voltage, have a lower leakage than any of the small-signal diodes.

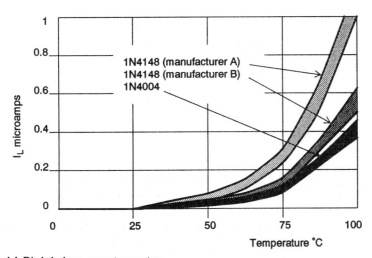

Figure 4.4 Diode leakage versus temperature

4.1.4   High-frequency performance

Up to now the diode characteristics discussed have been those which apply at d.c. As the frequency of use increases, a.c. characteristics become more important. The parameters of greatest interest are the equivalent capacitance, and the turn-on/turn-off behaviour.

The equivalent diode circuit includes a parallel capacitance. This is due to the depletion layer across the junction acting much as a dielectric separating two plates. As the applied reverse voltage changes, so does the width of the depletion layer, and so does the effective capacitance (Figure 4.5). A low reverse voltage gives a thin depletion layer and a high capacitance; a high reverse voltage reduces the capacitance.

This effect is exploited in the so-called "varicap" diode for voltage control of tuned circuits, used in virtually every modern radio receiver, and also in the varactor diode, where the non-linearity of the C/V law is used to generate rf harmonics. In most other

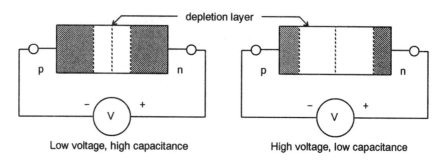

Figure 4.5 Diode junction capacitance versus applied voltage

applications it is either irrelevant or a nuisance. Actual capacitance depends on diode construction and varies from a few pF to hundreds of pF. If the circuit calls for high-speed or high-frequency operation then some allowance must be made for it in the design. Signal-switching applications, where the signal is small compared to the reverse biasing voltage, can assume a constant capacitance which can be reduced if necessary by increasing the bias voltage, other factors being equal. Large-signal or rectifying applications need to take into account the non-linearity of the capacitance/voltage relationship which most often manifests as unexpected waveform distortions.

The foregoing applies to diode capacitance under reverse bias. When the diode is forward biased the capacitance increases but the impedance is now low enough for this not to be a dominant issue.

4.1.5   Switching times

On turn-on, when reverse bias is changed to forward bias, the applied forward current has first to discharge the junction capacitance before the junction will conduct. Thus there is a delay in establishing the steady-state V_F, known as the forward recovery time, but no other adverse effects. Providing the diode is driven at turn-on from a reasonably low impedance, the turn-on time is much less than the turn-off time.

On turn-off, the applied reverse voltage must "sweep" all the conducting minority carriers out of the junction before conduction can cease, and this takes a finite time, during which current continues. At the end of this period the current drops to the expected reverse leakage value. Since the mechanism is quite different, there is no direct correlation between this time and the junction capacitance. "Reverse recovery" time is directly related to the forward current before reverse voltage is applied, and to the rate of change of current at turn-off. Figure 4.6 illustrates recovery times.

Reverse recovery

Reverse recovery time becomes an increasing embarrassment as switching speed and power increase, because it represents dissipated power at quite a high level ($V_R \cdot I$). The faster the switching frequency, the greater the proportion of power that is dissipated in the reverse direction; in high-power circuits this becomes a limiting factor on the diode rating, especially at the higher voltages, and also contributes to inefficiency in power conversion. Conventional rectifiers have recovery times in the $1 - 20\mu s$ region. To overcome this problem the "fast recovery" diode was developed, which by suitable processing reduces the reverse recovery time to a minimum, though not to zero. Typical recovery times are $150 - 200ns$ and fast recovery diodes are used extensively in high-

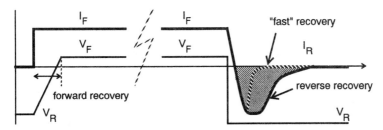

Figure 4.6 Diode forward and reverse recovery

voltage, high-speed switching. When even these speeds are too slow "ultra-fast" diodes are also available which can have recovery times down to 35 – 40ns.

Interference due to fast recovery

Fast recovery brings its own problems, though. The characteristic tail on turn-off "snaps" back very quickly to I_R, producing a very high transient rate of change of current (di/dt). Usually it is the highest di/dt within the circuit and so is responsible for most of the unwanted electromagnetic interference output. To save on the need for extra components to limit this current, yet another class of diodes have been developed, the "soft recovery" diodes, in which a compromise has been reached between speed of recovery and a comparatively gentle turn-off characteristic.

However, bear in mind that all p-n junction diodes exhibit some form of reverse recovery and are therefore capable of generating interference at harmonics of the switching frequency (even mains!) and of dissipating some power during this period.

4.1.6   Schottky diodes

The p-n semiconductor junction is not the only arrangement to show rectifying properties. A metal-semiconductor junction also rectifies. Devices which use this property are known as Schottky diodes. The important differences between conventional p-n silicon diodes and silicon Schottky diodes can be summarised as shown in Table 4.1.

Schottky diodes are used primarily for their low forward voltage or for their high

Conventional ▶︎◀	Schottky ▶︎◀
Forward voltage typically 0.6V at medium currents	Forward voltage typically 0.4V at medium currents
Minority carrier charge storage effects limit speed	No minority carriers, no charge storage, high speed
High reverse breakdown voltage achievable, in excess of 1kV	Low reverse breakdown voltage, no more than 50 - 100V typical
Low reverse leakage current	Higher reverse leakage current

Table 4.1 Schottky versus conventional diodes

speed. Available types are characterised for three main areas:

- high speed switching and general purpose
- rf and microwave mixers
- high-efficiency rectifiers

General purpose

Small signal Schottky diodes can be used in many of the same applications as conventional diodes as well as those where their lower V_F or high switching speed are essential. The shape of the V/I characteristic is the same though the values of V_F and V_{BR} differ. The temperature coefficient of V_F varies rather more with I_F, and is around -1mV/°C at the milliamp level. Leakage is up to an order of magnitude higher than typical p-n junctions and shows the same exponential temperature dependence. Schottky diodes are more expensive and less widely sourced than their conventional counterparts and this has restricted their widespread application; there is no Schottky equivalent to the universally popular 1N4148, which sells for less than 1p in quantity.

RF mixers

For rf applications, the Schottky diode is the almost ideal component for a mixer circuit, in which a deliberate non-linearity is introduced in order to extract the sum or difference of two frequencies applied to its inputs. The high speed, low noise and large signal handling ability of the Schottky make it particularly suitable for wideband mixers. The earliest applications for them were in this field and there is a wide range of devices characterised for such use.

Rectifiers

The largest growth area for Schottky rectifiers has been in the output stages of switch-mode power supplies. There is an enormous market for medium-to-high current 5V-output switchers for supplying computer circuits, and the trend is towards ever-greater efficiencies and higher switching speeds, for which the Schottky is eminently suited. At higher currents the forward voltage of a conventional rectifier can approach 1V, so that 20% of the total power of a 5V switcher is lost in the diode alone; the Schottky V_F of 0.5V cuts this to 10%. At the same time, the lack of a reverse recovery mechanism makes it designer-friendly at high speeds, and the low V_{BR} limitation is no hindrance at such a low output voltage. The higher unit cost of the Schottky is offset by the reduction in component count that can be achieved.

4.1.7   Zener diodes

By suitable selection of dimensions and impurities within the silicon it is possible to control the voltage at which reverse breakdown occurs. The slope of the diode V/I curve becomes quite flat in this region and the device can be used as a voltage regulator or clamp, and devices characterised for this purpose are called Zener diodes. The breakdown voltage can be controlled from 2.4V up to hundreds of volts, 270V being a practical maximum. In the forward direction the zener functions just like an ordinary silicon diode, with a somewhat higher V_F and an un-specified V_F/I_F characteristic.

Just as with other components, the zener is not perfect. Its slope resistance is not zero, its breakdown "knee" is not sharply defined, it has a leakage current below breakdown, and its breakdown voltage has tolerance and temperature coefficient. Figure 4.7 demonstrates these features.

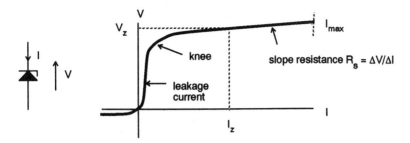

Figure 4.7 Zener reverse breakdown voltage-current characteristic

Slope resistance

Zeners are supplied for a quoted voltage, which is always defined at a given reverse current I_Z. At this current it will be within the specified tolerance, but at other currents it will differ, the difference being a function of the zener slope resistance R_s. The actual range of working voltage can be calculated by adding $(I - I_Z) \cdot R_s$ to the quoted voltage range, where I is the working current and I_Z is the current at which the zener voltage is quoted.

Over some range of I_Z, which you can determine from the published curves, R_s can be assumed to be fairly linear. As the current decreases the characteristic approaches the "knee" of the curve and R_s increases sharply. There is very little point in operating a zener intentionally on the knee. The actual knee current depends on the type and voltage but is rarely less than a few hundred microamps. Consequently, zeners are not much use for micropower or high-impedance circuits. For shunt voltage regulator applications at low currents, circuits based on the wide range of bandgap reference devices (see chapter 7) are preferable.

In a typical shunt stabiliser circuit (Figure 4.8) the voltage regulation is directly related to R_s. Clearly, the lower R_s the better. Slope resistance falls to a minimum around 6.8V and increases markedly at greater or lesser voltages. The lower voltage zeners have a much higher slope resistance than do those in the mid-range (Figure 4.9). Below 5V and above 100V the simple zener shunt stabiliser exhibits poor voltage regulation as a consequence. If a high-voltage zener is needed, better performance can be had by putting two or more lower-voltage devices in series to obtain the required voltage.

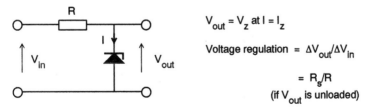

$$V_{out} = V_z \text{ at } I = I_z$$

$$\text{Voltage regulation} = \Delta V_{out}/\Delta V_{in}$$

$$= R_s/R$$
(if V_{out} is unloaded)

Figure 4.8 Regulation of a shunt stabiliser

Leakage

Below the knee, when the reverse voltage is not sufficient to begin the breakdown, there is still some current flow. This is due to leakage in the same way, and with the same temperature dependence, as a conventional diode. Leakage is usually specified for a zener at some voltage below the breakdown voltage, of the order of 20 - 30% less. It is an important specification when the zener is used as a clamp, so that the device's normal operating voltage is less than the breakdown value. A typical application is in protecting a circuit input from transient or continuous overvoltages (section 4.1.8).

Temperature coefficient

Like all components, the zener's breakdown voltage exhibits a temperature coefficient; but the zener tempco is somewhat more subtle than usual. There are in fact two mechanisms for reverse breakdown in silicon. Electron tunnelling is the dominant mechanism at low voltages and very thin junction barriers, while avalanche breakdown is dominant for higher voltages and thicker barriers. Depending on the required voltage, one mechanism or the other will predominate, and the crossover is at around 5V. The practical significance is that the two mechanisms have opposite temperature coefficients. They are also the reason for the dramatic variations in slope resistance. The optimum zener voltage for minimum temperature coefficient is between 4.7V and 5.6V, and where you have a choice of regulation voltage it is best to go for one of these values if tempco is important.

The graphs shown in Figure 4.9 illustrate the tempco and slope resistance variability for the Philips BZX79 range of zeners. Because these characteristics depend on the

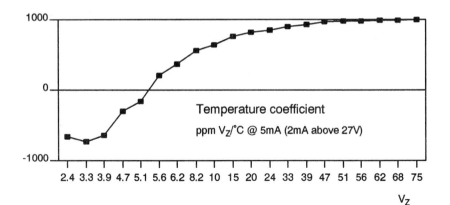

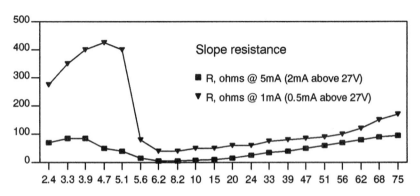

Figure 4.9 Zener slope resistances and temperature coefficients versus zener voltage for BZX79 series
Source: Philips Components published data

basic physics of the zener effect, other manufacturers' ranges will show similar performance.

Precision zeners

A further quirk of the process means that a device with a breakdown voltage of about 5.6 – 5.9V has a tempco of roughly +2mV/°C, which balances the tempco of a conventional forward biased silicon junction. By putting the two in series a virtually-zero-tempco zener can be created, with an effective breakdown voltage of between 6.2V and 6.4V. These are available off-the-shelf as "precision reference diodes" (the 1N821 series is the most common example) with a closely adjusted tempco and tolerance for use as voltage references. They are expensive, with the cost increasing in direct proportion to the tightness of specification. A similar effect can be achieved at around 8.4V, by putting two junctions in series with a 7.5V zener, which has a positive tempco of 4mV/°C. These devices compete directly with bandgap reference ICs in most aspects of performance and price; usually the bandgap wins out due to its lower slope resistance, lower operating current and more acceptable regulating voltage.

Zener noise

Another feature of zener breakdown is that it is a noisy process, electrically speaking. In fact, a zener operated at a constant current, ac-coupled and amplified is a good source of wideband white noise for calibration and measurement purposes. Zeners are not normally characterised for noise output so it is difficult to base a production design on them, but they can be used on a one-off basis. Noise is not usually a problem in voltage regulator applications, since it is many orders of magnitude below the dc zener voltage and can be virtually removed by the addition of a parallel decoupling capacitor. If the capacitor is omitted, either inadvertently or because a fast response is needed, then zener noise can be significant for precision references.

4.1.8   The Zener as a clamp

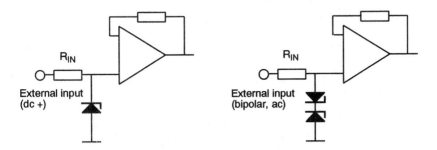

Figure 4.10 Input clamping by zener

In this application (Figure 4.10) the zener is used to prevent input overvoltages from damaging subsequent circuitry. It must not affect the input for voltages within the normal operating range but must clamp the op-amp input to a safe value when the external input is overloaded. The imperfections discussed in section 4.1.7 mean that it is not very good at doing this. Leakage current and capacitance restrict its use to comparatively low impedance and low frequency circuits. The breakdown voltage knee, tolerance and slope resistance mean that a practical limiting voltage must be substantially higher than the operating voltage and so the op-amp's full permissible input range is not utilised.

An application example

Take the bipolar-input circuit of Figure 4.10, assume that the op-amp has ±15V supplies and that its input pins are not allowed to exceed the supply voltage under possible fault conditions of 100V continuous input. The input itself can swing from +10V to –10V under normal operation. A back-to-back zener pair is used for clamping. The input source impedance is 10KΩ and the accuracy required of the voltage follower is 0.1% over a temperature range of 0 – 50°C. The zeners must be cheap and easily available.

The absolute maximum clamping voltage must be $(15V - V_F)$ less the zener tolerance. We can take V_F to be 0.8V ignoring tempco and slope resistance (in defiance of the previous strictures about diode characteristics!) and if the tolerance is 5%, which is the most readily available, V_z should be a maximum of 13.5V. The closest standard value is 13V, but this does not allow for voltage rise with power dissipation and slope resistance, which will exceed 0.5V. If we specify a BZX79 C12 (with a V_z of 12V), then from the published curves the maximum of 13.5V will be reached at around 25mA. This is within the BZX79's dissipation rating of 400mW at 50°C. This allows us to set the input resistor to be

$$R_{IN} = (100 - 13.5 - 0.8)/25mA = 3.4K\Omega, \text{ the nearest value being 3K3.}$$

Now assess the inaccuracy caused by diode leakage. The total input source resistance is 13.3KΩ and this is allowed to cause a 0.1% error on 10V, i.e. 10mV, due to leakage. This allows us a total leakage current of 10mV/13.3K = 0.75μA. Neglecting op-amp bias current (which may or may not be realistic) we can assign all this to the 12V zener operating at 10V reverse voltage. The data book gives us 0.1μA at 25°C and 8V, so doubling the leakage current for every 10°C rise in temperature we can expect 0.56μA at 50°C. We are just (*only* just!) within spec, and this has involved an act of faith that the leakage current at 10V for a lowest-tolerance zener will only be marginally worse than the quoted maximum at 8V. Because the higher-voltage zeners have a reasonably sharp knee we shall probably get away with it.

The diode capacitance of about 30pF at 10V (this is rarely specified, so you will probably have to guess or measure it) puts the 3dB roll-off of the input network at 400kHz. At lower voltages the capacitance increases and the bandwidth is correspondingly reduced. If the circuit is expected to operate up to these frequencies this will be a limiting factor on the use of a zener clamp.

However, despite its limitations the zener is a cheap, one- or two-component solution to input protection. Devices characterised especially as transient absorbers are also available, though not so cheap, and can be designed-in in the same manner when high energy transients are expected. Another approach is to incorporate diodes between the input and the supply rails. This is discussed in section 6.2.3.

4.2 Thyristors and triacs

The thyristor, or silicon controlled rectifier (SCR), is a four-layer diode whose conduction in the forward direction can be initiated by a trigger pulse or voltage at a third terminal called the gate. The triac is a similar device whose conduction can be initiated in both forward and reverse directions.

This class of devices (which includes further variants such as the diac, the

Rating	2-4A 600V	8A 600V	25A 600V
Thyristor	0.27	0.46	1.50
Triac	0.32	0.46	1.43

Typical prices: £, 100-up qty; plastic package devices

Table 4.2 Comparative costs of thyristors and triacs

unijunction transistor and the gate-turn-off thyristor) is extremely useful for power conversion and control and can be found in many other niche applications. Both its utility and its limitations stem from the fact that once conduction has been initiated, it continues until the applied current has been removed. If a thyristor is used in a dc circuit, once triggered it will remain conducting. If it is used in an ac circuit, it stops conducting at every zero-crossing and must be re-triggered on every half cycle.

4.2.1 Thyristor versus triac

Thyristors are used extensively both in light current applications and high-power switching and control, but triacs generally only find application in low power (below 40A) mains circuits. A triac costs much the same as a thyristor of a similar rating (Table 4.2) and is equivalent to two back-to-back thyristors, but this apparent advantage, in terms of component count, is also its greatest limitation. Since a triac conducts in both directions, it has only a brief interval during which sine-wave alternating current is passing through zero to recover to its blocked state. With inductive loads, the phase shift between current and voltage means that when the current falls to zero and the triac stops conducting, there will be an applied voltage across it. If this appears too rapidly, the triac will carry on conducting and control will be lost. Reliable operation is therefore limited to mains line frequencies and lower, and even at these frequencies snubbing (see section 4.2.6) is necessary for inductive loads.

As well as the characteristics that have already been discussed for conventional diodes – forward voltage and current, reverse breakdown voltage and leakage current, reverse recovery time – thyristors and triacs have another set of characteristics to consider, to do with triggering and conduction. These are trigger voltage and current, holding current and dV/dt.

The basic properties of silicon are the same for thyristors as for ordinary diodes. However, the thyristor construction is a pnpn sandwich between its main terminals, so the forward voltage drop is higher than that of an ordinary diode, generally from 0.8 to 2V depending on current. This restricts the thyristor's usefulness in low-voltage

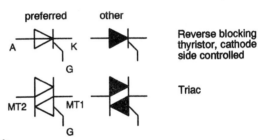

These are the accepted symbols (from BS3939/ IEC617) for the most common type of thyristor and triac.

preferred   other

Reverse blocking thyristor, cathode side controlled

Triac

Figure 4.11 Thyristor and triac symbols

circuits. Reverse breakdown and leakage mechanisms are the same and so the reverse characteristics are similar.

4.2.2   Triggering characteristics

Conduction is normally initiated by injecting energy into the gate terminal. The gate-cathode or gate-MT1 connection is a p-n junction and so is best driven from a low-impedance current source. Since triggering is energy-dependent a high current pulse can be applied for a short period or a lower current can be applied for longer. If the total energy content of the trigger pulse is not sufficient then unreliable triggering will occur, particularly at the lower-temperature extreme of operation when the energy required is greater. The typical behaviour of minimum required gate trigger current with variations in temperature and pulse width are shown in Figure 4.12.

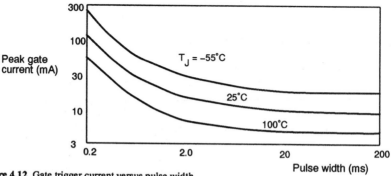

Figure 4.12 Gate trigger current versus pulse width

Thyristors can only be triggered by one polarity of current applied to the gate terminal. Triacs, on the other hand, can be triggered on either polarity of main terminal voltage by either polarity of trigger current. This is known as "four-quadrant" triggering and is a useful property as it means that a unipolar pulse can trigger both positive and negative half-cycles of the ac waveform. The triac's construction makes for different gate sensitivities in different quadrants; usually quadrant IV (Figure 4.13) is considerably less sensitive than the rest. A negative-going gate pulse, where possible, is preferable for equivalent triggering sensitivities under both main terminal polarities.

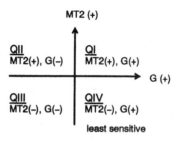

Figure 4.13 Triac triggering quadrants

4.2.3   False triggering

False triggering occurs when a spurious triggering pulse is coupled to the gate with sufficient amplitude to switch the device on, or if the main terminal (anode-cathode) blocking voltage is exceeded. It does not matter how conduction is initiated: once the device is conducting, it will continue to do so until the forward current is removed. The most usual coupling mechanism (Figure 4.14) is through the main-terminal-to-gate capacitance when a high rate-of-change of blocked voltage is present.

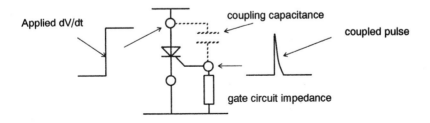

Figure 4.14 False triggering by coupled pulse to gate

The current coupled into the gate circuit can be calculated from $I = C \cdot dV/dt$, though you will often have to guess at C. There are two measures which you should take to guard against spurious coupling:

- prevent high dV/dt, if necessary by snubber circuits (see later). Thyristor and triac data sheets include a maximum dV/dt specification which should be observed.
- reduce the gate input impedance by means of a low-value parallel resistor, or even a capacitor. This calls for increased gate drive. So-called "sensitive gate" devices, with low gate drive requirements, are more susceptible to spurious triggering.

Driving the wanted trigger pulse via a pulse transformer directly to the gate with no other gate components is bad practice, because the leakage inductance of the transformer can present a high impedance to dV/dt coupled pulses; a low-value parallel resistor is still advisable. Noise coupled into the gate drive circuit can also cause false triggering. This can be a particular problem where the device is remote from its driver. Low gate impedance and/or a capacitor, and good wiring layout (see chapter 1) will reduce susceptibility. Overdrive the gate as far as is practicable (within dissipation limits) for the most reliable triggering.

4.2.4   Conduction

Once the device is triggered it will stay in conduction until the current through it drops to the holding current, I_H. The value of I_H is given in data sheets but is dependent both on temperature and gate impedance. Even for small thyristors this current can be quite high, of the order of several milliamps to tens of mA, and it represents a limit on the minimum load which can be switched. Clearly, if in ac applications a lightly-loaded device is triggered with a short pulse near the beginning of the half-cycle, the conduction current may not have built up to I_H before the trigger pulse ends (Figure 4.15). The device will then not conduct for that half-cycle. A longer duration trigger pulse will overcome this, but you may also prefer not to attempt to use the first few tens of degrees of conduction angle, especially since the sine-wave power in this part of the waveform is minimal.

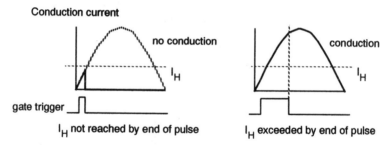

Figure 4.15 Effect of thyristor holding current

Reverse voltage on the gate increases the I_H value significantly, while forward bias will reduce it since the data sheet values are normally quoted for the gate open. Failure to appreciate this can cause latch and hold problems when thyristors are driven directly from a transistor, whose saturation voltage may reach hundreds of millivolts.

4.2.5 Switching

At turn-on, the rate of change of forward current (di/dt) should be limited. When the device is triggered the conduction region spreads relatively slowly through the silicon. If the turn-on current rises too rapidly then a high current flow is concentrated into a small region near the cathode, causing localised overheating and eventual device destruction. Maximum di/dt is sometimes specified and can be met by incorporating a small amount of inductance (calculated from L = –V/di/dt) into the load circuit; often the load itself is inductive enough for this. Permissible di/dt is strongly influenced by gate drive level and risetime, since a higher level of gate drive will spread the conduction region faster through the device. Higher gate drive also reduces the delay time from application of drive to turn-on, which is typically 1 – 2µs.

Turn-off

A reverse voltage cannot be used to turn off a triac, which conducts in both directions. However, thyristors *can* be turned off by reverse voltage, and their turn-off time has two components, reverse recovery time and forward blocking recovery time. The former has the same mechanism as a reverse-biased diode, i.e. removal of minority carriers from the reverse blocking junction with application of reverse voltage. The longer time constant is associated with forward blocking recovery and is the time required for the charge stored in the forward blocking junction to recombine. The total turn-off time is of the order of tens of microseconds, and is increased by increasing junction temperature and on-state current. Negative gate bias will decrease it, as this will speed up the removal of charge from the forward blocking junction.

4.2.6 Snubbing

Restrictions on dV/dt can be met by connecting a capacitor-resistor-diode network in parallel with the device. This technique is known as "snubbing", and it can apply to any switching circuit, not just to thyristor/triac circuits (earlier examples have been given for switching inductive circuits). The basic circuit is shown in Figure 4.16.

The rate-of-rise of turnoff voltage is determined by the time constant $R_L \cdot C$. R_L is the circuit minimum load resistance, for instance the cold resistance of a heater or lamp, the

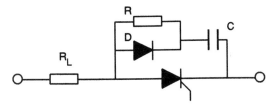

Figure 4.16 The thyristor snubber circuit

winding resistance of a motor or the primary resistance of a transformer. The resistor R limits the surge current through the device at turn-on due to discharge of C, and the diode D removes the influence of R while the applied voltage is rising. If the calculated value of R is of the same order of magnitude as or less than R_L then the diode can be omitted. In triac circuits when a diode is required, the entire snubbing circuit can be put into a diode rectifier bridge across the triac.

Values for R and C

C is calculated from

$$C = 0.63 \cdot V_{peak} / (dv/dt) \cdot R_L$$

where dv/dt is the device maximum specification and V_{peak} is the maximum voltage to which it will be exposed, e.g. for 240V phase-control applications V_{peak} will be 340V (although it would be prudent to allow a higher value if frequent transient spikes are expected on the supply).

R is calculated from

$$R = [V_{peak} / 0.5(I_{TSM} - I_L)]$$

or $\quad [V_{peak} / (C \cdot di/dt)]^{0.5}$

whichever is the larger, where I_{TSM} is the device half-cycle surge current rating, I_L is the maximum load current and di/dt is the rate-of-rise of current rating if it is quoted. The factor 0.5 in the first equation is a safety factor. Remember to check the resistor's power rating, allowing for pulse derating. The diode should have the same voltage rating as the triac or thyristor, but the half-cycle surge current rating need only be two or three times I_L as it only conducts for a short period at each turn-off.

A triac is controlling a 1kW cartridge heater at 240V. Assume the heater cold resistance is roughly one-tenth its hot resistance, which translates to a R_L of 6Ω and a peak turn-on current of 56A. The triac could be a TIC226M, which has a non-inductive dv/dt rating of 500V/μs and an I_{TSM} of 80A. This gives a C of

$$[0.63 \cdot 340 / 500 \cdot 6] = 0.07\mu F,$$

so use the next highest value of 0.1μF. R can be calculated from the first equation to be

$$[340 / (80 - 56) \cdot 0.5] = 28.3\Omega$$

Using a 27Ω resistor would give a di/dt of 4.7 A/μs. Because R is significantly higher than R_L, use of a parallel diode and putting the snubber in a bridge would be advisable.

Note that the selection of triac can have a large effect on the required values of the snubber components. For instance, a larger device, though unnecessary from the strict applications point of view, would have a larger I_{TSM} and could therefore get away with a lower R, which in turn might obviate the need for a diode. However, this might increase the stress on the capacitor which would need to be a pulse-rated device in any case. Alternatively, other triacs of similar rating may have an order of magnitude less dv/dt specification, which would need a much larger (and more expensive) capacitor.

4.3   Bipolar transistors

Much of what has been said about the characteristics of silicon diodes (see section 4.1) applies also to silicon transistors. Because the underlying mechanisms are the same, the forward and reverse conduction and high-frequency characteristics of the p-n junctions in either bipolar or field effect transistors will be the same as for the straightforward diode. The rest of this chapter will look at further characteristics which are peculiar to each type of device. Three families of transistor will be considered: bipolar, junction field-effect and MOS field-effect.

4.3.1   Leakage

Leakage current is as much of a problem in transistors as it is in the diode. It is normally specified in transistor data sheets as I_{CBO}, collector cutoff current (the collector-base current with emitter open circuit). This ensures that the specification ignores collector current due to amplified base current. It is particularly important in dc-coupled amplifier circuits, especially when collector leakage current in one transistor is injected into the base of the next and amplified.

A simple leakage example

Consider the simple two-transistor non-inverting buffer shown in Figure 4.18. This basic configuration is used both in switching circuits and in linear amplifiers. As a digital level shifter it can be used to interface between logic circuits and high voltage devices such as relays or stepper motors.

In the realisation of the circuit shown at Figure 4.18(a) TR1's entire collector current flows into TR2's base. This is fine when TR1 is fully conducting; the base current of TR2 is

$$I_B = (V+ - V_{BE2} - V_{CEsat1}) / R1$$

and this can easily be set to turn TR2 fully on and apply V+ to the load. When TR1 is

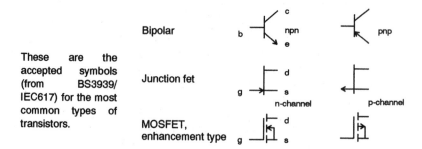

These are the accepted symbols (from BS3939/ IEC617) for the most common types of transistors.

Figure 4.17 Transistor symbols

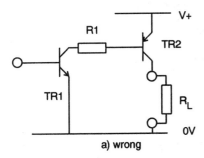
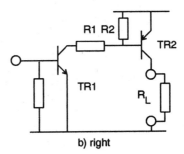

Figure 4.18 Biasing a two-transistor buffer

off it is a different matter. Even with its base shorted to ground its collector leakage current is all injected into TR2's base and thence amplified into R_L. At high temperatures the collector leakage can reach several microamps, or milliamps for a high-power device. If TR2 has a gain of a couple of hundred, then upwards of a milliamp could flow in R_L even when it is supposed to be off.

Worse, many applications will not tie the base of TR1 to ground when in the off state. An offset of even a few tens of millivolts will allow a small base current to flow which will be amplified and quickly swamp the collector leakage. If, say, a base-emitter junction passes 1mA at 600mV, it will pass 2.5µA at 100mV (from the diode equation); if TR1 has a gain of 100 and TR2 a gain of 200, then the current through R_L will be a respectable 50mA, quite significant for the off-state!

Adding a base-emitter resistor

The simple solution, shown in Figure 4.18(b), is to add a base-emitter resistor to any transistor which is threatened by leakage currents. The resistor is sized to divert only a modest proportion of the base current (typically one-tenth) when the transistor is being driven on. In the example above, assume that the base current of TR2 is set to 1mA in the on state; then taking $V_{BE} = 0.6V$ at this current, R_L is 0.6V / 0.1mA = 6kΩ. Now, if TR1 collector leakage reaches 10µA, this will develop no more than 60mV across TR2's base-emitter junction. From the diode equation again, only 45nA of this will be diverted into the base, which will not give any significant off-state leakage into R_L. Similar design can be applied to the resistor across TR1's base, though this depends on knowledge of the output characteristic of whatever drive circuit is used.

4.3.2   Saturation

Collector-emitter saturation voltage V_{CEsat} is the voltage which remains across the collector and emitter terminals when excess base current is applied to turn the device fully on. It is predominantly ohmic at higher collector currents, depending on the bulk silicon resistance between the terminals, but there is a residual voltage of between 50 – 200mV even at low collector currents which cannot be eliminated. V_{CEsat} is normally only specified at one or two values of collector and base current although data sheets will often give a graph of V_{CEsat} versus collector current. Increasing base current will only reduce V_{CEsat} marginally; setting it to more than a tenth of the operating collector current is pointless.

When a common-emitter switching transistor drives the base of another transistor, its saturation voltage may be enough, especially if it is operated at high power, to keep the second transistor partially on. The cure for this, as in the previous section, is to divide the voltage at TR2's base with another base-emitter resistor (Figure 4.19).

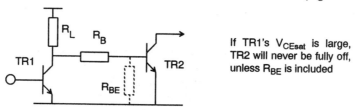

If TR1's V_{CEsat} is large, TR2 will never be fully off, unless R_{BE} is included

Figure 4.19 Minimising the effect of collector saturation voltage

The collector saturation voltage becomes significant in very low voltage circuits, when it represents a major fraction of the overall available collector voltage, and in high-current switching circuits, when it represents an appreciable power loss. It can also be a problem when the absolute value of the "on" level in switching circuits must be fixed. The temperature coefficient of V_{CEsat} is fairly low, generally less than 0.5mV/°C, but it has a complex dependence on collector current and junction temperature.

4.3.3   The Darlington

High saturation voltage is one particular disadvantage of the Darlington transistor. Darlingtons are essentially a transistor pair in a single package configured as in Figure 4.20 so as to multiply the current gain of each, so that overall gains of over 1,000 are readily achievable.

When it is driven into saturation the total voltage across the output transistor TR_B is the sum of TR_A's V_{CEsat} and TR_B's V_{BE}, because TR_A has to supply TR_B's base current. Normal operating V_{CEsat} of the Darlington is around 1V.

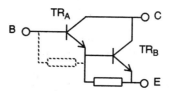

Figure 4.20 The Darlington transistor configuration

Total base-emitter voltage, of course, is double that of a conventional transistor because the base-emitter junctions are in series. The internal base-emitter resistors are not present in all types of device; they represent a trade-off between high gain, high switching speed and thermal stability. No base-emitter resistors means that all the input current for each stage flows into the base, so the current gain is maximised. But this increases susceptibility to thermal variations in leakage and V_{BE}, and it lengthens the switching turn-off time. Devices characterised for power switching use will normally include comparatively low-value base-emitter resistors.

4.3.4   Safe operating area

The safe operating area of a bipolar transistor is determined by four limits:

- maximum collector current
- maximum collector-emitter voltage
- maximum power dissipation
- second breakdown

The first three are always defined in the data for any transistor but the fourth is normally only applied to power transistors. Power dissipation ($I_C \cdot V_C + I_B \cdot V_B$) is specified for a given ambient or package temperature, usually 25°C, and must be reduced (de-rated) at higher temperatures using the quoted thermal resistance figures. Use a heatsink (section 9.5.2) to reduce the thermal resistance to ambient: do not expect a power transistor to achieve its maximum rated dissipation without one! Be careful even with small-signal devices if you are running them near their rated power at high ambient temperatures.

Second breakdown

Second breakdown is a phenomenon peculiar to bipolar devices which limits power dissipation further at high collector voltages. It is a thermal effect. If the transistor chip is thought of as a large number of elements in parallel, some of these will have a lower base forward voltage drop than others. Current will tend to concentrate in these, raising their temperature and lowering their voltage drop further. This will concentrate the current more, leading to local overheating and eventually a molten patch of silicon forming a short circuit between collector and emitter. Because it is a localised effect, it is independent of average junction temperature.

SOA curve

These four limits form the boundaries of the safe operating area for any transistor, and most manufacturers will provide a plot of the form shown in Figure 4.21 for their power transistors. The second breakdown limit normally intersects the power dissipation limit at some point, although for small-signal devices the locus of second breakdown may lie outside the safe operating area altogether.

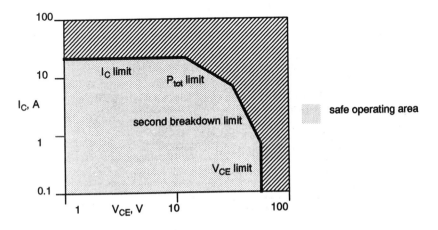

Figure 4.21 Typical curve of safe operating area

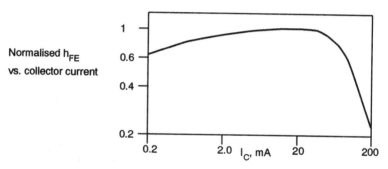

Figure 4.22 Typical curve of gain versus current

4.3.5 Gain

Current gain is not an entirely straightforward transistor parameter. It is to be found in data sheets under the heading h_{FE}; note that this refers to dc current gain and is different from h_{fe}, which is small-signal ac current gain. It is normally specified between a minimum and a maximum, and for some devices graded versions are available which have a tighter specification band. For example, the popular BC548 has a published h_{FE} band from 110 to 800 but is available in A, B and C selections which have bands from 110 – 220, 200 – 450 and 420 – 800 respectively. In many applications a very wide h_{FE} range is a real design headache, and in some instances – the BC548 is one – there is no cost penalty in selecting a graded device.

Current gain varies with collector current, voltage and temperature. Most data sheets will present a curve of the form shown in Figure 4.22 for gain versus current. Each transistor is optimised for a particular range of operating current and gain drops off substantially either side of this range, more so at the higher current end. When considering gain in the circuit design remember to allow for operating current if this is different from that at which the gain is specified. Similarly for collector-emitter voltage; lower voltages than quoted will cause a more dramatic fall-off in gain at the upper end of the operating current range. This is sometimes the cause of waveform distortion, as depicted in Figure 4.23, due to apparent lack of drive in large-signal transistor amplifiers. There may in fact be enough base drive current for the expected gain, but the combination of high collector current and low collector voltage reduces the gain so far that it is insufficient to fully turn-on the device.

Higher temperature increases the gain while lower temperature reduces it, by up to a factor of 2 – 3 for the widest temperature extremes. In fact, transistor gain is such a variable quantity – depending on the individual device, on operating conditions and on temperature – that no design should ever rely on it as a parameter for fixing circuit operation. Instead, seek to minimise the effect of its variations, and also remember that over at least some parts of the circuit's operating envelope, gain will be less than the minimum you find in the data sheet.

4.3.6 Switching and high frequency performance

The section on diodes (section 4.1.5) has already shown how the p-n junction turn-off differs from turn-on due to the need to sweep the minority carriers out of the junction. The same principle applies to transistor switching. Bipolar transistors have longer turn-off times than turn-on: the actual times can be modified by the transistor's construction.

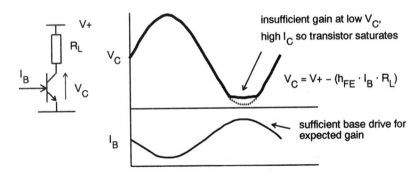

Figure 4.23 Waveform distortion due to gain fall-off

A typical small-signal switching transistor might have a turn-on time of less than 50ns and a turn-off of 100 to 200ns whereas a general-purpose amplifier device can be several times slower, and its data sheet will not include switching time figures. Do not use a general-purpose type (such as the BC54... series) and expect fast switching. The trade-off is between switching speed and gain.

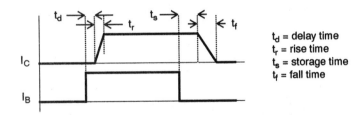

Figure 4.24 Transistor switching times

Data sheets generally quote four switching time figures: delay, rise, storage and fall times. Add the first two to get the turn-on time and the last two to get turn-off. The relationships are shown in Figure 4.24.

Individual manufacturers may specify their switching times slightly differently. Also, check the switching test circuit: storage and delay times can be reduced by overdriving the base heavily and the test circuit is usually quite different from the circuit that you will be using, so treat the quoted figures with caution. Turning off the base with a large negative voltage is common in test circuits, and is a good way to speed up the turn-off, but is often difficult to implement with the constraints of a real circuit. You should also bear in mind that reverse V_{BE} breakdown voltage is low, usually between 7 and 10V. Rise, fall and delay times all fall with increasing collector current.

A major disadvantage of the Darlington compared to conventional bipolars is its low switching speed. This is because its configuration worsens the bipolar's already poor turn-off time. To improve storage time, and hence turn-off, it is necessary either to prevent saturation, or to provide a way to reverse the direction of base current on turn-off; the Darlington can do neither.

Speeding up the turn-off

Transistor switching is conventionally taken to mean between fully-off and fully-on (saturated). If you prevent the device from saturating then the storage time, which is due to excess base current, is reduced to zero. This can be done in one of two ways: use an emitter-coupled pair as the basic switch, or divert the base current through a base-collector Schottky diode (Figure 4.25). The first is the principle behind the emitter-

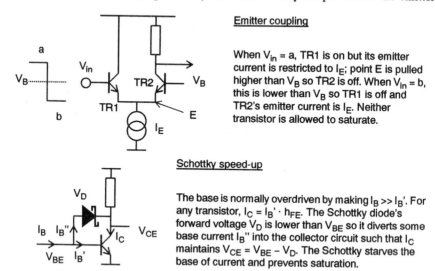

Emitter coupling

When V_{in} = a, TR1 is on but its emitter current is restricted to I_E; point E is pulled higher than V_B so TR2 is off. When V_{in} = b, this is lower than V_B so TR1 is off and TR2's emitter current is I_E. Neither transistor is allowed to saturate.

Schottky speed-up

The base is normally overdriven by making $I_B \gg I_B'$. For any transistor, $I_C = I_B' \cdot h_{FE}$. The Schottky diode's forward voltage V_D is lower than V_{BE} so it diverts some base current I_B'' into the collector circuit such that I_C maintains $V_{CE} = V_{BE} - V_D$. The Schottky starves the base of current and prevents saturation.

Figure 4.25 Speed-up methods for transistor switches

coupled logic (ECL) IC family, and the second is the principle behind Schottky/low-power Schottky TTL (S/LSTTL) logic. The second technique can be applied less effectively with a conventional diode; this circuit arrangement is known as the "Baker clamp". Both techniques give much faster switching for a given collector current.

4.3.7   Grading

In talking about gain, we have already touched on the subject of transistors selected and marked for a specific range of one particular parameter. In fact, this is the way that most transistors are made. A given transistor die may end up bearing any of a whole variety of different part numbers, depending on how the manufacturer tests and characterises it. Grading for gain is one aspect; the same applies when testing collector breakdown voltage, so that for example a BC547 and a BC548 can be from the same batch but one has been found to break down at a higher voltage than the other. Similarly devices may be graded for noise, so that a BC549 is from the same batch as a BC548 but has passed the noise test. Alternatively, transistors of a quite different part number may be taken from the same batch but have parameters tested to much closer limits than the base type, and these can then be sold at a premium. By this means a transistor manufacturer can maximise the yield from any batch of devices, merely by marking and characterising them differently.

The other side of this is that cost, and sometimes availability, of a device depends on how tightly it is specified. This is not always the case, as for example in the gain and voltage grades of the BC54... series, which generally all come out at around the same cost because they are produced in such high volume that yields of any particular grade can be optimised at will. But the best approach is to use the part with the most relaxed specification that is acceptable – or even no spec at all if it is for a parameter that doesn't matter, as for instance may be noise – as this will be the easiest and the cheapest to source.

4.4 Junction Field Effect transistors

The field effect transistor (FET) differs from the bipolar device in that current is transported by majority carriers only, whereas in the bipolar carriers of both polarities (majority and minority) are involved. For this reason it was originally known as the "unipolar" transistor. FETs are classified according to the mechanism used by their control terminal as junction FETs (JFETs) or metal-oxide-semiconductor FETs (MOSFETs). Figure 4.26 compares the basic transfer functions of p- and n-channel JFETs and MOSFETs.

The JFET consists of a channel of semiconductor material along which majority carrier current may flow. The current is controlled by a voltage applied to a reverse-biased p-n junction (the gate) formed along the channel. Because the gate is reverse-biased virtually no current flows in it. Therefore, in contrast to the bipolar which is a low-impedance, current-controlled device, the JFET is a high-impedance, voltage-controlled device. This characteristic is shared with the thermionic valve (vacuum tube), and circuit designs for one can often be transposed to the other with no change other than operating voltage.

The two ends of the channel are connected to the source and drain terminals. In practice, though the JFET symbol is shown as asymmetrical the channel geometry is usually symmetrical and it does not matter which terminal is used as the drain and which as the source, since the channel will conduct equally well in either direction. Devices for some applications, notably rf amplifiers, are constructed asymmetrically to optimise inter-electrode capacitances and should not be reverse connected.

4.4.1 Pinch-off

JFETs work in the depletion mode, which means that current conduction is controlled by depletion of the carriers within some region of the channel. This depletion can be brought about either by increasing the reverse-biased gate voltage or by increasing the drain-source voltage, resulting in a family of output characteristic curves as shown in Figure 4.27.

There are two distinct regions of operation, depending on applied drain-source voltage. With $V_{GS} = 0V$, i.e. gate shorted to source, the drain current increases linearly with voltage and the channel acts like a pure resistor, until the pinch-off voltage V_p is reached. At this point the drain current I_D saturates at I_{DSS} and beyond it current is essentially constant and independent of V_{DS}. As the gate voltage is increased the saturation region extends towards zero and the saturated drain current reduces, until $V_{GS(off)}$, the gate-source cutoff voltage, is reached at which point I_D approaches zero. Because the same physical mechanism is in play, the magnitudes of V_p and $V_{GS(off)}$ are the same, though they are of opposite polarity.

A casual glance through a few JFET data sheets will show that $V_{GS(off)}$ can vary from unit to unit over a very wide range, six-to-one being not uncommon. This is one of the major disadvantages of the device as it greatly complicates bias design, especially

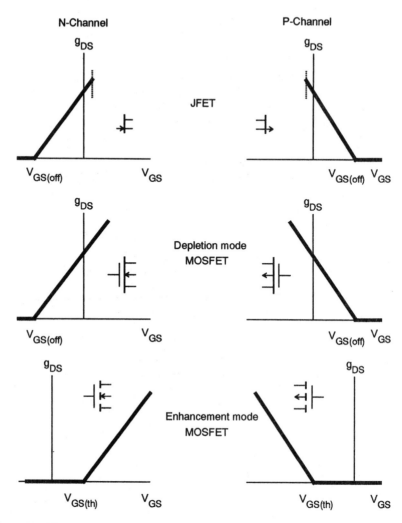

Figure 4.26 FET channel conductance g_{DS} versus gate-source voltage V_{GS}

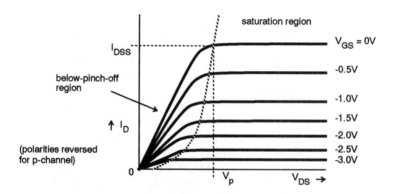

Figure 4.27 Typical family of output characteristics for n-channel JFET

for low-voltage applications which cannot afford the extra supply voltage needed to accommodate it.

4.4.2   Applications

Use of JFETs incurs a cost penalty. The cheapest general purpose JFET starts at around 20p in medium quantity, compared to around 5p for general purpose bipolars. Thus JFETs are restricted to applications which can use their special characteristics. Of these the foremost are

- analogue switches
- rf amplifiers, mixers and oscillators
- constant current regulators
- high input impedance amplifiers

Analogue switches

The virtue of the FET for use as an analogue switch is that its channel is purely resistive. There is no input-output offset voltage, and leakage currents from the control terminal to input or output are very small and often negligible. The input signal can be of either polarity but is limited by the available gate switching voltage and by the gate-channel breakdown rating. Because the JFET operates in depletion mode (Figure 4.26 explains FET operating modes), the on-state requires that the gate is connected to the source while the off-state requires a gate voltage that is more negative than the source by at least $V_{GS(off)}$ (or more positive, for a p-channel device). This means that the driver supply voltage must be greater by several volts than the expected signal input range. Also, the gate drive circuitry cannot be a straightforward logic output, as it must follow the analogue signal during the on-state. A typical switch-plus-driver circuit is as shown in Figure 4.28.

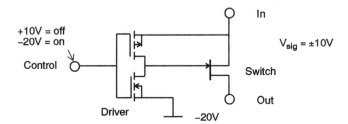

Figure 4.28 Analogue switch configuration

Because of the complications of the driver circuit it is easier and cheaper to use integrated circuit analogue switches, where the FET is integrated along with its driver circuit, than to use a discrete FET and separate driver. This offers the extra benefit that important circuit characteristics such as control signal feedthrough and variation in "on" resistance are already characterized. You should only need to use discrete FET circuits when operating outside the voltage range of readily available IC switches.

RF Circuits

The JFET has been popular as a small-signal rf device (amplifier, mixer, oscillator) for many years. It has low rf noise, good inter-electrode capacitance stability and well defined rf input and output impedances, which make it easier to design with than bipolars. In addition, its square-law transfer characteristic allows a high dynamic range. Many devices characterised for hf and vhf use are available at reasonable prices, of the order of 20 – 50p for plastic packages.

Current regulators

A frequent niche application for JFETs is as a one-component current regulator. If the device is connected with gate shorted to source (Figure 4.29) then once V_{DS} exceeds the pinch-off voltage the drain current will limit to I_{DSS} and remain constant over a wide range of applied voltage. The current can be adjusted downwards from I_{DSS} by including a resistor in series with the source, which effectively gives a constant gate bias.

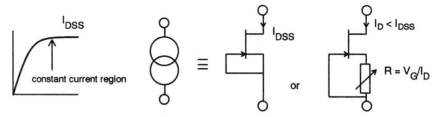

Figure 4.29 The JFET as current regulator

The circuit cannot be used as a precision current source/sink because its output impedance is on the low side for such applications, and the control current is temperature dependent, though like the zener there is a zero-tempco crossover point. Also, the wide variation in pinch-off voltage and I_{DSS} between devices results in a similarly wide variation of current. Even so, it can be useful where the absolute value of current is unimportant, such as in amplifier biasing. It is possible to obtain "current regulator diodes" which are gate-source-shorted JFETs specially characterised for the purpose and available in reasonably tightly selected current bands from 0.2 to 5mA, though these are relatively expensive, being over £1 for the higher performance parts and around 50p for low-performance plastic types.

4.4.3   High impedance circuits

The JFET is very useful for the design of high input impedance amplifiers. Because the gate under normal operating conditions is a reverse-biased junction, the low-frequency input impedance of a JFET front end is limited only by gate leakage and by the resistance of any bias resistor that may be necessary. Gate leakage currents of a few picoamps at room temperature are readily achievable, although making use of them is another matter – leakage currents due to stray paths across pcbs and connectors are usually the limiting factor. The JFET does not always live up to its promise of high impedance, though.

Firstly, because the current is due to reverse-biased junction leakage, it increases exponentially with temperature in the same way as was seen for the ordinary silicon diode junction. Thus at the maximum commercial temperature limit of 70°C the leakage current is more than twenty times that at room temperature, though it is still better than that of most other devices. At 125°C, the maximum military temperature limit, it is a thousand times worse, and a well-designed bipolar input will have a better performance. This is one reason why JFET input circuits are rare in military designs.

The gate current breakpoint

Secondly the common-mode voltage range is restricted. Obviously if the input voltage exceeds the supply rails then incorrect operation may occur – though you might want to exploit the region between zero V_{GS} and $V_{GS(off)}$ in order to exceed the supply voltage in one direction or the other. A more subtle mechanism also affects input impedance, which is that gate leakage current is critically dependent on drain-gate voltage and drain current. Gate leakage current is normally characterized as I_{GSS}, i.e. gate current with drain shorted to source. In an operational circuit, an n-channel JFET behaves differently.

N-channel JFETs experience an I_G "breakpoint" above which the gate current rises rapidly with increasing drain-gate voltage. This results from a phenomenon similar to avalanche breakdown but modified by the availability of carriers in the channel due to drain current. These carriers are accelerated by the drain-gate electric field and contribute a leakage current through the gate because of their ionization on impact with the silicon. Consequently, the breakpoint value depends on I_D and is between a third to a half of drain-gate breakdown voltage, as can be seen from Figure 4.30.

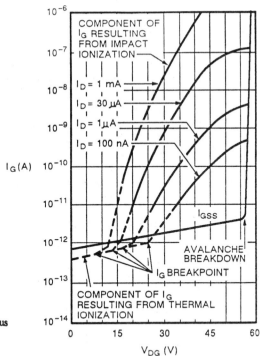

Figure 4.30 Plot of gate current versus drain-gate voltage for a typical JFET
Source: Siliconix

Depressed Z_{in}

The effect of this is that the input impedance of a JFET amplifier will be lower, perhaps by several orders of magnitude, than that calculated from the I_{GSS} figure once the I_G breakpoint is exceeded. This places a limit on the effective common-mode input voltage range. Note that p-channel JFETs, because of their lower carrier mobility, do not exhibit the effect to anything like the same degree.

In order to overcome the problem the drain-gate bias voltage must be held below the breakpoint. A reduction in drain current will shift the breakpoint knee higher, but the improvement is only marginal. In simple source-follower circuits it may be possible to reduce the drain bias voltage sufficiently as in Figure 4.31(a), but if the full operating

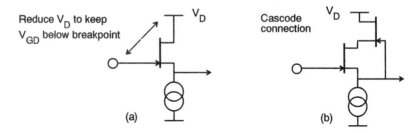

Figure 4.31 Reducing the effect of gate current breakpoint

range of the fet is required then a better solution is to add a second fet in cascode with the first (b). The drain-gate voltage of the input fet is maintained at a constant low voltage over the whole input range by the source of the cascode fet. The latter is still subject to operation above the breakpoint at the negative end of the input voltage range, but the excess gate current is now added to the operating current with negligible effect.

4.5 MOSFETs

The distinguishing feature of the MOSFET, or metal oxide semiconductor FET, is that its gate is insulated from the semiconducting channel by a thin layer of oxide. It is not a p-n junction so, unlike the JFET, no gate leakage current flows.

4.5.1 Low-power MOSFETs

Low-power MOSFETs have existed for some time, aimed at niche applications similar to JFETs, particularly high input impedance amplifiers, rf circuits and analogue switches. These devices are planar fabricated, with the source and drain terminals diffused into one side of the substrate. Development of the planar MOSFET into the double-diffused (DMOS) structure has allowed smaller channel dimensions which result in lower capacitances and faster switching times. DMOS devices have useable performance as analogue switches in the sub-nanosecond range, and as rf amplifiers into the GHz range, but they are not widely sourced.

Gate breakdown

The drawback of low-power MOSFETs centres around the extremely thin insulating layer between gate and channel. Generally manufacturers specify a breakdown voltage for the gate of between ±15V and ±40V. At the same time the gate-channel capacitance is a few pF, and since there is no discharge path for this capacitance when the device is out of circuit the charge needed to exceed the breakdown voltage is very small (around a hundred pico-coulombs) and can easily be generated electrostatically (*cf* section 9.2.2). As a result low-power MOSFETs can be destroyed simply by handling them prior to insertion into the circuit.

There are two solutions to this problem. One is to impose rigorous anti-static handling precautions at all stages of production. The safest procedure is to ensure that all leads of the device are shorted together, for example by wrapping a wire link around them, until the circuit board soldering is complete. This is labour-intensive compared to most other assembly operations, is prone to error and is also impossible to achieve with surface mount packages. Unprotected-gate MOSFETs are therefore not popular for high volume production.

Protection for the gate

The other solution is to prevent the gate voltage from reaching the breakdown level by incorporating extra protection, in the form of a zener diode rated to just below the breakdown voltage (Figure 4.32), within the package of the device. Manufacturers will

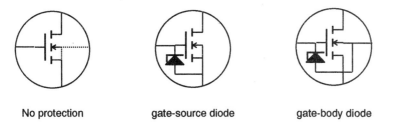

No protection   gate-source diode   gate-body diode

Figure 4.32 MOSFET gate protection

often offer versions of the same device with and without gate zener protection. The great advantage of including protection is that the device can now be treated in the assembly process just like any other semiconductor component, and no additional assembly precautions need be taken.

The trade-off is in the circuit performance, of course. The zener restricts negative gate voltage swings (for an n-channel device) to one diode drop, which can affect the design of the gate drive circuit. More importantly, it adds a component of gate leakage current which exhibits all the properties of diode leakage outlined earlier. In some applications this can nullify the advantage of using a MOSFET. Compare, for instance, the published characteristics for the Siliconix SD210 and SD211. The SD210 has no protection diode and a quoted maximum I_{GBS} gate leakage current of 0.1nA. The SD211 has an integral gate zener and a maximum I_{GBS} of 10μA (the typical figure being 10^{-6} lower!). There is no other difference in performance except for breakdown voltages: the devices use the same chip.

MOSFET tradeoffs

The choice facing designers who want to use low-power MOSFETs is fairly simple. Either specify a gate-protected device and accept the performance limitations that this implies, but make yourself popular with the production department. Or, insist on using unprotected devices because you need ultra-high impedance, and impose extra costs in production, along with possible reliability penalties if the production procedures are not rigorously enforced.

Remember also that the susceptibility of low-power MOSFETs to gate breakdown may not end after they are soldered into circuit. Especially if they are used in high-impedance input circuits, the gate can still be vulnerable if its biasing resistor is in the high megohm range (or worse still, absent) and handling precautions may need to be

	Power MOSFET	Power Bipolar
Max voltage rating	Typically 500V	Typically 700V
Max current rating	Up to 70A	Up to 100A
Switching speeds	Typically better than 100ns, independent of temperature	$0.3 - 5\mu s$ typical
Input characteristics	Voltage, $5 - 10V\ V_{GSth}$ gain tempco $-0.2\%/^\circ C$	Current, $20 - 100\ h_{FE}$ h_{FE} tempco $+0.8\%/^\circ C$
Output characteristics	Resistive, current sharing when paralleled, R_{DSon} tempco $0.7\%/^\circ C$	Non-resistive, current hogging when paralleled V_{CEsat} tempco $0.25\%/^\circ C$
Breakdown	No second breakdown Safe operating area is thermally limited Static-induced breakdown precautions advisable	Second breakdown at high V_{CE} Susceptible to thermal runaway
Unit Cost[†]	Average 2.5p/watt with 10:1 variation	Average 1p/watt for conventional, 0.6p/watt for darlington

† Average cost per watt of quoted maximum power dissipation for plastic packaged devices, npn or n-channel, 80 - 100V voltage rating, > 5W power rating

Table 4.3 Comparison of power MOSFET and Bipolar characteristics

observed at the board or equipment level.

4.5.2   VMOS Power FETs

The constraint of channel on-resistance (typically tens to hundreds of ohms) and consequent restriction on power handling of the conventional planar MOSFET was removed with the development of the double-diffused vertical MOSFET or VMOS. This range of devices, introduced by International Rectifier in the late 70's under the HEXFET trademark and subsequently widely second-sourced, can achieve "on" resistances in the milliohm region, and drain-source breakdown voltages up to 500V and higher. It is therefore a direct competitor to the power bipolar transistor and outperforms it in many respects. Table 4.3 compares the major differences between the two. The necessary tradeoffs in switching speed, gain and power handling that bipolars impose are easier to handle or even absent with VMOS.

Although there are many clear performance advantages the VMOS's cost penalty, on average two-and-a-half times more expensive per watt, will often rule it out of cost-sensitive applications. Comparing cost-per-watt is slightly misleading since bipolar devices may have to be derated to 60-70% of their nominal power rating at high V_{CE} because of second breakdown considerations, which is not necessary for VMOS. Also, newer VMOS parts continue to be introduced at lower costs and it is likely that some devices will soon claim cost parity with their bipolar counterparts.

The VMOS's features versus bipolar for power switching applications are generally claimed to be

- voltage drive rather than current drive
- increased switching speeds
- resistive output characteristic

Each of these although real, also brings with it disadvantages, and requires some effort to properly realise in practice.

4.5.3   Gate drive impedance

Because the VMOS is a majority carrier device, it does not have any of the minority-carrier switching delays associated with bipolars and its switching speed is governed by how fast the gate can be driven. Since the gate impedance is almost entirely capacitive, the resulting turn-on/turn-off delays depend on the output impedance of the driver circuit (Figure 4.33).

Thus, although it is possible to drive a VMOS switch directly from a CMOS logic gate, doing so will compromise the achievable switching time. For instance, a typical 74HC-series gate at 5V has an output drive capability of 4mA; a typical medium-size VMOS may have an input capacitance of 200pF. If the switching voltage threshold of the VMOS is 3V, then the time taken to reach this level is $C \cdot V/I = 150ns$. Note that most

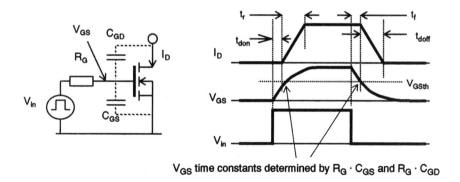

V_{GS} time constants determined by $R_G \cdot C_{GS}$ and $R_G \cdot C_{GD}$

Figure 4.33 Gate switching capacitance and waveforms

VMOS devices are characterised for R_{DSon} at $V_{GS} = 10V$, and should not be driven directly from 5V logic. Worst-case combinations of supply voltage, output level and VMOS threshold will lead to the VMOS being driven on the knee of its "on" characteristic, with unpredictable and high values of R_{DSon}. Families of "logic-compatible" VMOS are now available, characterised at $V_{GS} = 5V$. However, this is by no means the end of the story; the actual switching waveforms are complicated by feedback of the drain waveform through the drain-gate capacitance C_{GD}. This also has to be charged by the gate drive circuit, along with C_{GS}. Keeping R_G low improves switching times significantly. Various techniques can be employed to this end; Figure 4.34 shows some of them.

Gate-source overvoltage

Besides switching times, other factors call for a low dynamic impedance from the gate drive circuit. Excessive V_{GS} will punch through the gate-source oxide layer and cause

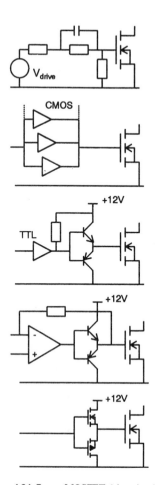

Cheap & cheerful:
gate speed-up capacitance, used with
excess drive voltage V_{drive}

With CMOS buffers, the more the
merrier. Drive from the highest
available voltage

TTL should be interfaced with a high-
voltage open-collector buffer, pull-up
resistor and complementary emitter
followers

Complementary emitter followers can
also be used to speed-up a linear drive
circuit

The ultimate driver: a power CMOS
pair. Choose your driver output
resistance by selecting devices

Figure 4.34 Power MOSFET drive circuits

permanent damage. Transient gate-source overvoltages can be generated by large drain
voltage spikes (caused, for instance, by another device connected to the drain, or by
induced transients) coupled through the drain-gate capacitance. If the dynamic drive
circuit impedance is high – as might be the case if the gate is driven from a pulse
transformer – the transient amplitude will only be limited by the potential divider effect
of C_{GD} and C_{GS}. A typical ratio of these values is 1:6, so that a 300V drain transient
would be reflected as a 50V gate transient, which is quite enough to destroy the gate.

The simple precaution if such transients are anticipated (especially if the drain is
connected to an external circuit) is to specify a device with an integral zener gate
protection diode, or to incorporate one in the external circuit near to the gate-source
terminals. VMOS power devices, because of their higher gate capacitance, are
inherently less susceptible to handling-induced electrostatic damage than are low-
power MOSFETs, but it is nevertheless prudent to specify gate-protected parts if they
are available.

4.5.4   Switching speed

High switching speeds, while desirable for minimizing power losses, have two highly undesirable side effects. They generate significant electromagnetic interference (EMI) and they also generate drain-source overvoltage spikes. EMI is primarily affected by layout, filtering and screening and is the subject of chapter 8. Here, it is merely noted as a factor opposing the trend to faster switching.

Drain-source overvoltages are generated when current flowing through an inductive load is switched off fast (*cf* section 3.4.4). Even when the main load is non-inductive or

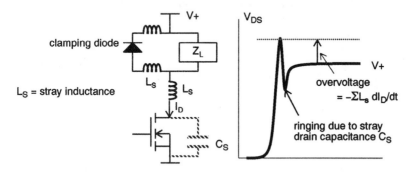

Figure 4.35 Drain-source turn off transient

is clamped, stray inductance can cause significant transients at the current levels and switching speeds offered by VMOS devices (Figure 4.35). For instance, 20A being switched across 0.5μH in 50ns will create a spike of 200V. Additionally, the forward recovery of the clamping diode may be insufficient to catch the leading edge of the main inductive transient. As well as ensuring that all connections in the high-di/dt area are as short as possible, a drain-source clamping zener local to the device is a useful precaution.

4.5.5   On-state resistance

Finally, it is also wise to remember when running a VMOS device in the on-state at high power that dissipation is thermally limited. This means that I^2R must be restricted to maintain the junction temperature T_j below the maximum permitted in the device data sheet, usually 150°C. The section on thermal management (section 9.5) outlines how to relate these two factors. However on-state resistance, R_{DSon}, has a positive temperature coefficient and is normally quoted at a given temperature and V_{GS}. At maximum T_j it will be around 1.8 – 2 times the quoted value at 25°C so, if the heatsink is sized so that maximum expected dissipation gives maximum permitted temperature, allowable operating current will be about 0.7 times that available at 25°C. Also if steady-state applied V_{GS} less than that for which R_{DSon} is quoted, the actual R_{DSon} will be higher again. Data sheet figures should never be taken at face value.

The advantage of a positive tempco of R_{DSon} is that it makes paralleling of devices to achieve higher current much easier. With a negative tempco, if a particular device takes more than its share of current it gets hotter, and this reduces its resistance, causing it to take more current, and so on. This is known as current hogging and is a feature of bipolar transistors, leading to thermal runaway if not dealt with, usually by including a separate emitter resistor for each transistor. The VMOS positive tempco on the other hand encourages current sharing and prevents thermal runaway.

P-channel VMOS

Although the foregoing discussion has focussed on n-channel VMOS fets, the same considerations apply to p-channel devices. Many manufacturers offer "complementary" p-channel parts for their more popular n-channel ranges. Because of the differing resistivity of the two base materials there is no such thing as a truly complementary pair. P-type silicon has a much higher resistivity than n-type, so the p-channel device requires a larger active area to achieve the same on-resistance. This reflects in the cost: a p-channel part of the same R_{DSon} and voltage rating will be more expensive (typically three times) than its complementary n-type. Gate threshold voltage, transconductance and capacitances can be nearly equalized. However, thermal resistance, pulsed and continuous current rating, and safe operating area are all higher for the p-channel as a result of its larger die. This has the happy consequence that whenever matched operation is needed, the p-channel will have greater operating margins with respect to its thermal ratings.

Chapter 5

Linear integrated circuits

The operational amplifier is the basic building block for analogue circuits, and progress in op-amp performance is the "litmus test" for linear IC technology in much the same way as progress in memory devices is for digital technology. This chapter will be devoted to op-amps and comparators, with a tailpiece on voltage references. This is not to deny the enormous range of other analogue functions that are available, but these are intended for specific niche applications and little can be generalised about them.

Volumes have already been written about op-amp theory and circuit design and these aspects will not be repeated here. Rather, we shall take a look at the departures from the ideal op-amp parameters that are found in practical devices, and survey the tradeoffs – including cost and availability, as well as technical factors – that have to be made in real designs. Some instances of anomalous behaviour will also be examined.

5.1   The ideal op-amp

The following set of characteristics (in no particular order, since they are all unattainable) defines the ideal voltage gain block:

- infinite input impedance, no bias current
- zero output impedance
- arbitrarily large input and output voltage range
- arbitrarily small supply current
- infinite open-loop gain
- zero input offset voltage and current
- zero noise contribution
- absolute insensitivity to temperature, power rail and common-mode input fluctuations
- zero cost
- off-the-shelf availability in any package
- compatibility between different manufacturers
- perfect reliability

Since none of these features are achievable, you have to select a practical op-amp from the multitude of imperfect types on the market to suit a given application. Some basic examples of tradeoffs are

- a high-frequency ac amplifier will need maximum gain bandwidth but won't be interested in bias current or offset voltage

- a battery-powered circuit will want minimum supply current
- a consumer design will need to minimise the cost at the expense of technical performance
- a precision instrumentation amplifier will need minimum input offsets and noise but can sacrifice speed and cost

Device data sheets contain some but not all of the necessary information to make these tradeoffs (most crucially, they say nothing about cost and availability!). The functional characteristics often need some interpretation and critical parameters can be hidden or even absent. In general, if a particular parameter you are interested in is not given in the data sheet, it is safest to assume a pessimistic figure. It means that the manufacturer is not prepared to test his devices for that parameter or to certify a minimum or maximum value.

5.2   The practical op-amp

5.2.1   Offset voltage

Input offset voltage V_{OS} can be defined as that differential dc voltage required between the inverting and non-inverting inputs of an amplifier to drive its output to zero. In the perfect amplifier, zero volts in will give zero volts out; practical devices will show offsets ranging from tens of millivolts down to a few microvolts. The offset appears as an error voltage in series with the actual input voltage. Definitions vary, but a "precision" op-amp is usually considered to be one that has a V_{OS} of less than $200\mu V$ and a V_{OS} temperature coefficient (see later) of less than $2\mu V/°C$. Bipolar input op-amps are the best for very-low-offset voltage applications unless you are prepared to limit the bandwidth to a few tens of Hz, in which case the CMOS chopper-stabilized types come into their own. The chopper technique achieves very low values of V_{OS} and drift by repeatedly nulling the amplifier's actual V_{OS} several hundred times a second with the aid of charge storage capacitors.

Offsets are always quoted referenced to the input. The output offset voltage is the input offset times the closed-loop gain. This can have embarassing consequences particularly in high-gain ac amplifiers where the designer has neglected offset errors because, for performance purposes, they are unimportant. Consider a non-inverting ac-coupled amplifier with a gain of 1000 as depicted in Figure 5.1.

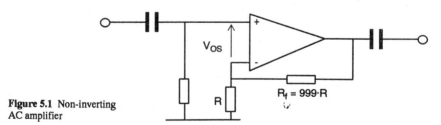

Figure 5.1 Non-inverting AC amplifier

Let's say the circuit is for audio applications and the op-amp is one half of a TL072 selected for low noise and wide bandwidth, running on supply voltages of ±12V. The TL072 has a maximum quoted V_{OS} of 10mV. In the circuit shown, this will be amplified by the closed loop gain to give a dc offset at the output of 10V – which is far too close to the supply rail to leave any headroom for large overload ac signals. In fact, the TL072 is likely to saturate at 9–10V anyway with ±12V power rails.

Output saturation due to amplified offset

The designer may be wanting 2mV pk-pk ac signals at the input to be amplified up to 2V pk-pk signals at the output. If the dc conditions are taken for granted then you might expect at least 20dB of headroom: ±1V output swing with ±10V available. But, with a worst-case V_{OS} device virtually no headroom will be available for one polarity of input and 20V will be available for the other. Unipolar clipping will result. The worst outcome is if the design is checked on the bench with a device which has a much-better-than-worst-case offset, say 1mV. Then the dc output voltage will only be 1V and virtually all the expected headroom will be available. If this design is let through to production then the scene is set for unexpected customer complaints of distortion! An additional problem presents itself if the output coupling capacitor is polarised: the dc output voltage can assume either polarity depending on the polarity of the offset. If this isn't recognised it can lead to early failure of the capacitor in some production units.

Reducing the effect of offset

The solutions are plentiful. The easiest is to change the feedback to ac-coupling which

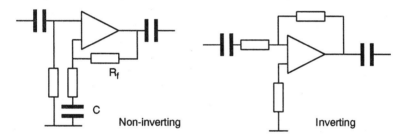

Figure 5.2 AC coupling to reduce offset

gives a dc gain of unity so that the output dc voltage offset is the same as the input offset (Figure 5.2). The inverting configuration is simpler in this respect. The difficulty with this solution is that the time constant $R_f \cdot C$ can be inordinately long, leading to power-on delays of several seconds.

The second solution is to reduce the gain to a sensible value and cascade gain blocks. For instance, two ac-coupled gain blocks with a gain of 33 each, cascaded, would have the same performance but the offsets would be easily manageable. The bandwidth would also be improved, along with the out-of-band roll-off, if this were necessary. Unfortunately, this solution adds components and therefore cost.

A third solution is to use an amplifier with a better V_{OS} specification. This will either involve a tradeoff in gain-bandwidth, power consumption or other parameters, or cost. For instance, in the above example PMI or Linear Technology's OP227 might be a suitable candidate, though it is noticeably more expensive. The overall cost might work out the same though, given the reduction in components over the second solution.

Offset drift

Offset voltage drift is closely related to initial offset voltage and is a measure of how V_{OS} changes with temperature and time. Most manufacturers will specify drift with temperature, but only those offering precision devices will specify drift over time.

Present technology for standard devices allows temperature coefficients of between 5 and 40μV/°C, with 10μV/°C being typical. For bipolar inputs, the magnitude of drift is directly related to the initial offset at room temperature. A rule of thumb is 3.3μV/°C for each millivolt of initial offset. This drift has to be added to the worst case offset voltage when calculating offset effects and can be significant when operating over a wide temperature range.

Early MOS-input op-amps suffered from poor offset voltage performance due to gate threshold voltage shifts with time, temperature and applied gate voltage. New processes, particularly developments in silicon gate technology, have overcome these problems and CMOS op-amps (Texas Instruments' LinCMOS™ range for instance) can achieve bipolar-level V_{OS} figures with extremely good drift, $1 - 2μV/°C$ being quoted.

Circuit techniques to remove the effect of drift

Microprocessor control has allowed new analogue techniques to be developed and one of these is the nulling of input amplifier offsets, as in Figure 5.3. With this technique the initial circuit offsets can be calibrated out of the system by applying a zero input, storing the resultant input value (which is the sum of the offsets) in non-volatile memory andd subsequently subtracting this from real-time input values. With this technique, only offset drifts, not absolute offset values, are important. Alternatively, for the cost of a few extra components – analogue switches and interfacing – the nulling can be done repetitively in real time and even the drift can be subtracted out. (This is the microprocessor equivalent of the chopper op-amps discussed earlier.)

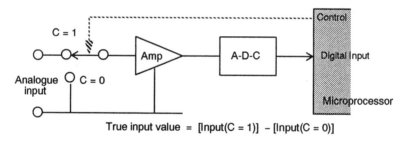

Figure 5.3 Offset nulling with a microprocessor

5.2.2   Bias and offset currents

Input bias current is the average dc current required by the inputs of the amplifier to establish correct bias conditions in the first stage. Input offset current is the difference in the bias current requirements of the two input terminals. A bipolar input stage requires a bias current which is directly related to the current flowing in the collector circuit, divided by the transistor gain. FET-input (or BiFET) op-amps on the other hand do not require a bias current as such, and their input currents are determined only by leakage and the need for input protection.

Bias current levels

Input bias currents of bipolar devices range from a few microamps down to a few nanoamps, with most industry-standard devices coming in at 0.5μA max. There is a well-established tradeoff between bias current and speed; high speeds require higher first-stage collector currents to charge the internal node capacitance faster, which in turn requires higher bias currents. Precision bipolar op-amps achieve less than 20nA while some devices using current nulling techniques can boast picoamp levels. JFET and CMOS devices routinely achieve input currents of a few picoamps or tens of picoamps at 25°C, but because this is almost entirely reverse-bias junction leakage it increases exponentially with temperature (see section 4.1.3). Industry standard JFET op-amps are therefore no better than bipolar ones at high temperatures, though precision JFET and CMOS still show nanoamp levels at the 125°C extreme. Note that even the 25°C figure for JFETs can be misleading, because it is quoted at 25°C *junction* temperature: many JFET op-amps take a fairly high supply current and warm up significantly in operation, so that the junction temperature is actually several degrees or tens of degrees higher than ambient.

The significance of input bias and offset currents is twofold: they determine the steady-state input impedance of the amplifier and they result in added voltage offsets. Input impedance is rarely quoted as a parameter on modern op-amp data sheets since bias currents are a better measure of actual effects. It is irrelevant for the closed-loop inverting configuration, since the actual impedance seen at the op-amp input terminals is reduced to near zero by feedback. The input impedance of the non-inverting configuration is determined by the change in input voltage divided by the change in bias current due to it.

Output offsets due to bias and offset currents

Of more importance is the bias current's contribution to offsets. The bias current flowing in the source resistance R_S at each terminal generates a voltage in series with the input; if the bias currents and source resistances were equal the voltages would cancel out and no extra offset would be added.

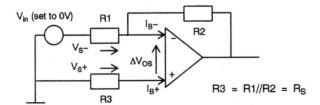

Ideal situation: I_B, R_S equal at + and − terminals, so $V_{S+} = V_{S-} = I_B \cdot R_S$ and $\Delta V_{OS} = 0$

Bad design: R_S not equal at + and − terminals so, neglecting I_{OS},
$V_{S+} = I_B \cdot R3$, $V_{S-} = I_B \cdot R1//R2$ and $\Delta V_{OS} = I_B \cdot (R1//R2 - R3)$

Practical op-amp: I_{B-} differs from I_{B+} by I_{OS}, R_S equal at both terminals, so
$V_{S+} = I_{B+} \cdot R_S$, $V_{S-} = (I_B + I_{OS}) \cdot R_S$ and $\Delta V_{OS} = I_{OS} \cdot R_S$

Figure 5.4 Bias and offset currents

As it is, the offset current generates an effective offset voltage given by $I_{OS} \cdot R_S$ (with a temperature coefficient determined by both) which adds to, or subtracts from, the inherent offset voltage V_{OS} of the op-amp. Clearly, whichever dominates the output depends on the magnitude of R_S. Higher values demand an op-amp with lower bias and offset currents. For instance, the current and voltage offsets generated by a 741's input circuit are equal when $R_S = 33K\Omega$ (typical $V_{OS} = 1mV$, $I_{OS} = 30nA$). The same value for the TL081 JFET op-amp is $1000M\Omega$ ($V_{OS} = 5mV$, $I_{OS} = 5pA$).

I_B itself does not contribute to offset *provided* that the source resistances are equal at each terminal. If they are not then the offset contribution is $I_B \cdot \Delta R_S$. Since I_B can be an order of magnitude higher than I_{OS} for bipolar op-amps, it pays to equalize R_S: this is the function of R3 in the circuit above. R3 can be omitted or changed in value if current offset is not calculated to be a problem. Apart from the disadvantage of an extra component, R3 is also an extra source of noise (generated by the noise component of I_B) which can weigh heavily against it in low-noise circuits.

5.2.3   Common mode effects

Two factors, which because they don't appear in op-amp circuit theory can be overlooked until late in the design, are common mode rejection ratio (CMRR) and power supply rejection ratio (PSRR). Figure 5.5 shows these schematically. Allied with these is common mode input voltage range.

CMRR

An ideal op-amp will not produce an output when both inputs, ignoring offsets, are at the same (common-mode) potential throughout the input range. In practice, gain differences between the two inputs, and variations in offset with common-mode voltage, combine to produce an error at the output as the common-mode voltage varies. This error is referred to the input (that is, divided by the gain) to produce an equivalent input common-mode error voltage. The ratio of this voltage to the actual common-mode input voltage is the common-mode rejection ratio (CMRR), usually expressed in dB. For example, a CMRR of 80dB would give an equivalent input voltage error of 100µV for every 1V change at both + and − inputs together. The inverting amplifier configuration is inherently immune to common-mode errors since the inputs are held at a constant level, whereas the non-inverting and differential configurations are susceptible.

CMRR is not necessarily a constant. It will vary with common-mode input level and temperature, and always worsens with increasing frequency. Individual manufacturers may specify an average or a worst-case value, and will always specify it at dc.

PSRR

Power supply rejection is similar to CMRR but relates to error voltages referred to the input as a result of changes in the power rail voltages. As before, a PSRR of 80dB with a rail voltage change of 1V would result in an equivalent input error of 100µV. Again, PSRR worsens with increasing frequency and may be only 20–30dB in the tens-to-hundreds of kiloHertz range, so that high-frequency noise on the power rails is easily reflected on the output. There may also be a difference of several tens of dB between

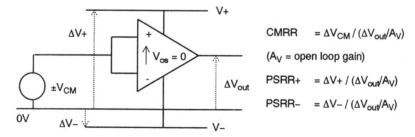

Figure 5.5 Common mode and power supply rejection ratio

the PSRRs of the positive and negative supply rails, due to the difference in internal biasing arrangements. For this reason it is unwise to expect equal but anti-phase power rail signals, such as mains frequency ripple, to cancel each other out.

5.2.4   Input voltage range

Common mode input voltage range is usually defined as the range of input voltages over which the quoted CMRR is met. Errors quickly increase as it is exceeded. The input range may or may not include the negative supply rail, depending on the type of input. The popular LM324 range and its derivatives have a pnp emitter coupled pair at the input, which allows operation down to slightly below the negative rail. The CMOS-input devices from RCA and Texas also allow operation down to the negative rail. All these op-amps stop a few volts short of the positive rail, as they are optimised for operation from a single positive supply. Conventional bipolar devices of the 741 type, designed for ±15V rails, cannot swing to within less than 2V of each rail, and BiFET types are even more restricted.

Absolute maximum input

The common-mode operating input voltage is different from the absolute maximum input voltage range, which is usually equal to the supply voltage. If you exceed the maximum input voltage without current limiting then you are likely to destroy the device; this can quite easily happen inadvertently, apart from circuits connected to external inputs, if for instance a large value capacitor is discharged directly through the input. Even if current is limited to a safe value, over-voltages on the input can lead to unpredictable behaviour. Latch-up, where the IC locks itself into a quasi-stable state and may draw large currents from the power supply, leading to burnout, is one possibility. Another is that the sign of the inputs may suddenly change, so that the inverting input suddenly becomes non-inverting. (This was a well known fault on early devices such as the 709.) These problems most frequently arise with capacitive coupling direct to one or other input, or when power rails to different parts of the circuit are turned on or off at different times.

5.2.5   Output parameters

Two factors constrain the output voltage available from an op-amp: the power rail voltage, and the load impedance.

Power rail voltage

It should be obvious that the output cannot swing to a greater value than either power rail. Unfortunately it is often easy to overlook this fact, particularly as the power connections are frequently omitted from circuit diagrams, and with different quad op-amp packages being supplied from different rails it is hard to keep track of which device is powered from what voltage. More seriously, with unregulated supplies the actual voltage may be noticeably less than the nominal. The required output must be calculated for the worst-case supply voltage.

Most op-amps cannot swing their output right up to either supply rail. The exceptions are the CMOS-output devices represented by RCA's CA3130 series, and many of the types intended for single-supply operation which have a current sink at the output and can reach within a few tens of millivolts of the negative (or ground) supply terminal. The rest of the conventional bipolar and biFET parts cannot swing to within less than 2V of either rail. The classic output stage (Figure 5.6) is a complementary

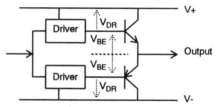

Figure 5.6 Output voltage swing restrictions

emitter follower pair which gives low output impedance, but the output available in either direction is limited by $(V_{DR(min)} + V_{BE})$. Depending on the detailed design of the output, the swing may or may not be symmetrical in either polarity. This fact is disguised in some data sheets where the maximum peak-to-peak output voltage swing is quoted, rather than the maximum output voltage relative to the supply terminals.

Load impedance

Output also depends on the circuit load impedance. This may again seem obvious, but this writer has more than once been assured, apparently seriously, that feedback reduces the output impedance of an op-amp in proportion to the ratio of open- to closed-loop gains and therefore it should be capable of driving very low load resistors. Well of course to an extent it is, but Ohm's law is not so easily flouted and a low output resistance is only driven with a low output voltage swing, depending entirely on the current drive capability of the output stage. The maximum output current that can be obtained from most devices is limited by package dissipation considerations to about ±10mA. In some cases, the output current spec is given as a particular output voltage swing when driving a stated value of load, typically $2 - 10K\Omega$. If you want more output current it is quite in order to buffer the output with an external complementary emitter follower or something similar, provided that feedback is taken from the final output. Take care with short-circuit protection when doing this (or else don't be surprised if you have to keep replacing transistors) and also bear in mind that you have changed the high frequency response of the combination and the closed-loop circuit may now be unstable.

Output current protection is universally provided in op-amps to prevent damage when driving a short circuit. This does not work in the reverse direction, that is when the output voltage is forced outside either supply rail by a fault condition. In this case there will be one or two forward-biased diode junctions to the power rail and current will flow through these limited only by the fault source impedance. Preventative measures for circuits where this is likely are dealt with in section 6.2.3.

5.2.6   AC parameters

The performance of an op-amp at high frequency is described by a motley collection of parameters, each of which refers to slightly different operating conditions. They are:

- large-signal bandwidth, or full-power response: the maximum frequency, at unity closed-loop gain, for which a sinusoidal input signal will produce full output at rated load without exceeding a given distortion level. This bandwidth figure is normally determined by the slew-rate performance.

- small-signal or unity-gain bandwidth, or gain-bandwidth product: the frequency at which the open-loop gain falls to unity (0dB). The "small-

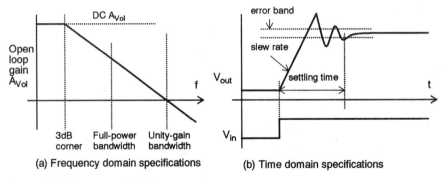

(a) Frequency domain specifications   (b) Time domain specifications

Figure 5.7 AC op-amp specifications

signal" label means that the output voltage swing is small enough that slew-rate limitations do not apply.

- slew rate: the maximum rate of change of output voltage for a large input step change, quoted in volts per microsecond.

- settling time: elapsed time from the application of a step input change to the point at which the output has entered and remained within a specified error band about the final steady-state value.

These parameters are illustrated in Figure 5.7.

5.2.7   Slew rate and large signal bandwidth

These two specifications are intimately related. All conventional op-amps can be modelled by a transconductance gain block driving a transimpedance amplifier with capacitive feedback (Figure 5.8).

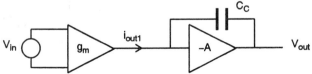

Figure 5.8 Op-amp slewing model

The compensation capacitor C_C is the dominant factor setting the op-amp's frequency response. It is necessary because a feedback circuit would be unstable if the gain block's high frequency response was not limited. Digital designers avoid capacitors in the signal path because they slow the response time, but this is the price you must pay for freedom from unwanted oscillations when working with linear circuits.

Slew rate

The exact value of the price is measured by the slew rate. From the above circuit, you can see that the rate of change of V_{out} is determined entirely by i_{out1} and C_C (remember $dV/dt = I/C$). As an example, the 741's input section current source can supply 20μA and its compensation capacitor is 30pF, so its maximum slew rate is 0.67V/μs. Op-amp designers have the freedom to set both these parameters within certain limits, and this is what distinguishes a fast, high-supply-current device from a slow, low-supply-current one. "Programmable" devices such as the LM4250 or LM346 make the trade-off more obvious by putting it in the circuit designer's hands.

If i_{out1} can be increased without affecting the transconductance g_m, then slew rate can be improved without a corresponding reduction in stability. This is one of the major virtues of the biFET range of op-amps. The JFET input stage can be run at high currents for a low g_m relative to the bipolar and so can provide an order of magnitude or more increase in slew rate.

Large-signal bandwidth

Slew-rate limitations on dV_{out}/dt can be equated to the maximum rate-of-change of a sinewave output. The time derivative of a sinewave is

$$d/dt\,[\,V_p\,\sin\omega t\,] \quad = \quad \omega{\cdot}V_p\,\cos\omega t \quad \text{where } \omega = 2\pi f$$

This has a maximum value of $2\pi f{\cdot}V_p$, which relates frequency directly to peak output voltage. If V_p is equated to the maximum dc output swing then f_{max} can be inferred from the slew rate and is equal to the large signal or full power bandwidth,

$$2\pi \cdot f_{max} \quad = \quad \text{slew rate} / V_p$$

Slewing distortion

Operating an op-amp above the slew-rate limit will cause slewing distortion on the output. In the limit the output will be a triangle wave (Figure 5.9) as it alternately switches between positive and negative slewing, which will decrease in amplitude as the frequency is raised further. If the positive and negative slew rates differ there will be asymmetrical distortion on the output. This can generate an unexpected effect equivalent to a dc offset voltage, due to rectification of the asymmetrical feedback waveform or overloading of the input stage by large distortion signals at the summing junction. Also, slewing is not always linear from start to finish but may exhibit a fast rise for the first part of the change followed by a reversion to the expected rate for the latter part.

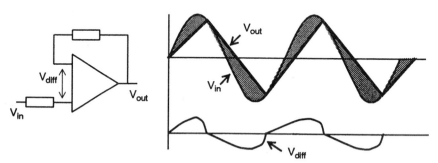

Figure 5.9 Slewing distortion

5.2.8   Small-signal bandwidth

The op-amp frequency response shown in Figure 5.7(a) exhibits the same characteristic as a simple low-pass RC filter. The 3dB frequency or corner frequency is that point at which the open-loop gain has dropped by 3dB from its dc value. It is set by the compensation capacitor C_C and is in the low Hertz or tens of Hertz range for most devices. The gain then "rolls off" at a constant rate of 20dB per decade (a ten-times increase in frequency produces a tenfold gain reduction) until at some higher frequency the gain has dropped to 1. This frequency therefore represents the unity-gain bandwidth of the part, also called the small-signal bandwidth.

The fact of a constant roll-off means that it is possible to speak of a constant "gain-bandwidth product" (GBW) for a device. The LM324's op-amps for instance have a typical unity-gain bandwidth of 1MHz, so if you wanted to use them at this frequency you could only use them as voltage followers – and small-signal ones at that, since large output swings would be slew-rate limited. A gain of 10 would be achievable up to 100kHz, a gain of 100 up to 10kHz and so on (but see the comments on open-loop gain later). This gain-bandwidth tradeoff is illustrated in Figure 5.10. On the other hand, the

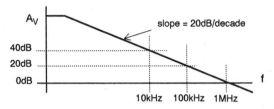

Figure 5.10 Gain-bandwidth roll-off

more recent OP-27 has a unity-gain bandwidth of 8MHz and can therefore offer reasonable gains up to the MHz region.

5.2.9   Settling time

When an op-amp is faced with a step input, as compared to a linear function such as a sinusoid or triangle wave, the step takes some time to propagate to the output. This time includes the delay to the onset of output slewing, the slewing time, recovery from slew-limited overload, and settling to within a given output error. Students of feedback theory will know that a feedback-controlled system's response to a step input exhibits some degree of overshoot (Figure 5.7(b)) or undershoot depending on its damping factor. Op-amps are no different. For circuits whose output must slew rapidly to a precise value, particularly analogue-to-digital converters and sample-and-hold buffers, the settling time is an important parameter.

Op-amps specifically intended for such applications include settling time parameters in their specifications. Most general-purpose ones do not, although a graph of output pulse response is often presented from which it can be inferred. When present, settling time is usually specified for unity gain, relatively low impedance levels, and low or no capacitive loading. Because it is determined by a combination of closed-loop amplifier characteristics both linear and non-linear, it cannot be directly predicted from the open-loop specs of slew rate and bandwidth, although it is reasonable to assume that an amplifier which performs well in these respects will also have a fast settling time.

5.2.10   Uncompensated op-amps

The presence of an internal compensating capacitor relieves you as circuit designer of a major burden, because you know that the op-amp is inherently stable no matter what value of gain is set by external feedback. This peace of mind is bought at the expense of the best high-speed performance of which the op-amp may be capable. In order that they may not be accused of stifling circuit designers' initiative, op-amp designers also offer another option: the uncompensated, or partially compensated, op-amp.

This sub-species, of which the 709 and the LM301A were the forerunners, omits the compensation capacitor (or reduces it) so that the device is no longer unity-gain-stable, but provides external pins for it so that you can add your own. This gives the ability to set speed and bandwidth characteristics by changing one component (Figure 5.11), although you now have to be more careful of the gain and phase characteristics of the feedback network at high frequencies – not always a simple matter, as these are precisely the frequencies at which unpredictable circuit strays become significant.

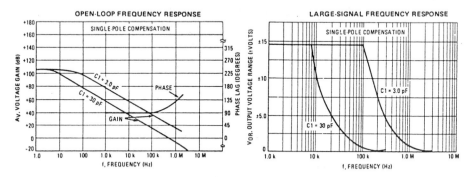

Figure 5.11 Open-loop small and large signal frequency response of the LM301A
Source: Motorola

Since the advent of relatively cheap high-performance high-speed op-amps such as the OP-27, the value of uncompensated devices to the average circuit designer has become more questionable (though the OP-27 has a partially-compensated brother, the OP-37, which boasts a unity-gain bandwidth of 63MHz). Initially the omission of an internal compensation capacitor allowed a smaller die size and thus a lower price, but this advantage has now eroded, and nowadays the major reason for most designers continuing to work with these devices is the reduction in stockholding that may be made possible by using one part for different applications – though this is often a very worthwhile advantage and should not be denigrated.

5.2.11   The oscillating amplifier

Just about every analogue designer has been bugged by the problem of the feedback amplifier that oscillates (and its converse, the oscillator that doesn't) at some time or other. There are really only a few fundamental causes of unwanted oscillations, they are all curable, and they can be listed as follows:

- feedback-loop instability
- incorrect grounding
- power supply coupling
- output stage instability
- parasitic coupling

The most important clue in tracking down instability is the frequency of oscillation. If this is near the unity-gain bandwidth of the device then you are most likely suffering feedback-induced instability. This can be checked by temporarily increasing the closed-loop gain. If feedback is the problem, then the oscillation should stop or at least decrease in frequency. If it doesn't, look elsewhere.

Feedback-loop instability is caused by too much feedback at or near the unity-gain frequency, where the op-amp's phase margin is approaching a critical value. (Many books on feedback circuit theory deal with the question of stability, gain and phase margin, using tools such as the Bode plot and the Nyquist diagram, so this isn't covered here.) Incorrect external compensation of an uncompensated op-amp is often a contributory factor.

Ground coupling

Ground loops or other types of incorrect grounding cause coupling from output back to input of the circuit via a common impedance in its grounded segment. This effect has been covered in chapter 1 but the circuit topology is repeated here, in Figure 5.12. If the

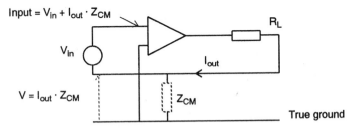

Figure 5.12 Common-impedance ground coupling

resulting feedback sense gives an output component in-phase with the input then positive feedback occurs, and if this overrides the intended negative feedback you will have oscillation. The frequency will depend on the phase contribution of the common impedance and can vary over a wide range.

Power supply coupling

Power supplies should be properly bypassed to avoid similar coupling through the common-mode power supply impedance. Power supply rejection ratio falls with frequency, and typical $.01 - .1\mu F$ capacitors may resonate with the parasitic inductance of long power leads in the MHz region, so these problems usually show up in the 1 – 10MHz range. Using $1 - 10\mu F$ tantalum capacitors for power rail bypassing will drop the resonant frequency and stray circuit Q to the level at which problems are unlikely (compare Figure 3.17 for capacitor resonances).

Output stage instability

Localized output-stage instability is most common when the device is driving a capacitive load. This can create output oscillations in the high-MHz range which are generally cured by good power-rail decoupling close to the power supply pins, with the decoupling ground point close to the return point of the load impedance, or by including a low-value series resistor in the output within the feedback loop.

Capacitive loads also cause a phase lag in the output voltage by acting in combination with the op-amp's open-loop output resistance (Figure 5.13). This increased phase shift reduces the phase margin of a feedback circuit. A typical capacitive load, often invisible to the designer because it is not treated as a component, is a length of coaxial cable. Until the length starts to approach a quarter-wavelength at the frequency of interest, coax looks like a capacitor: for instance, 10 metres of the popular RG58C/U 50Ω type will be about 1000pF. The capacitance can be decoupled from the output with a low-value series resistor, and high-frequency feedback provided by a small direct feedback capacitor C_F which compensates for the phase lag caused by C_L.

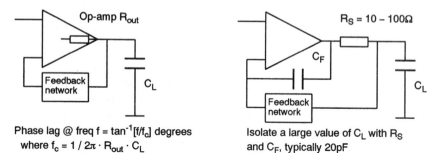

Phase lag @ freq f = tan⁻¹[f/f_c] degrees
where $f_c = 1 / 2\pi \cdot R_{out} \cdot C_L$

Isolate a large value of C_L with R_S and C_F, typically 20pF

Figure 5.13 Output capacitive loading

Stray capacitance at the input

A further phase lag is introduced by the stray capacitance C_S at the op-amp's inverting input. With normal layout practice this is of the order of 3 – 5pF which becomes significant when high-value feedback resistors are used, as is common with MOS- and JFET-input amplifiers. The roll-off frequency due to this capacitance is determined by the feedback network impedance as seen from the inverting input. The small-value direct feedback capacitance C_F of Figure 5.14 can be added to combat this roll-off, by roughly equating time constants in the feedback loop and across the input. In fact this technique is recommended for all low-frequency circuits as with it you can restrict loop bandwidth to the minimum necessary, thereby cutting down on noise, interference susceptibility and response instability.

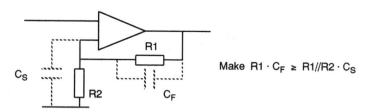

Make $R1 \cdot C_F \geq R1//R2 \cdot C_S$

Figure 5.14 Adding feedback capacitance

Parasitic feedback

Finally in the catalogue of instability sources, remember to watch out for parasitic coupling mechanisms, especially from the output to the non-inverting input. Any coupling here creates unwanted positive feedback. Layout is the most important factor: keep all feedback and input components close to the amplifier, separate input and output components, keep all pc tracks short and direct, and use a ground plane and/or shield tracks for sensitive circuits.

5.2.12   Open-loop gain

One of the major features of the classical feedback equation which is used in almost all op-amp design,

$$A_{CL} = A_{OL}/(1 + A_{OL} \cdot \beta)$$

where β is the feedback factor,
A_{OL} is the open-loop gain,
A_{CL} is the closed-loop gain

is that if you assume a very high A_{OL} then the closed-loop gain is almost entirely determined by β, the feedback factor. This is set by external (passive) components and can therefore be very tightly defined. Op-amps always offer a very high dc open-loop gain (80dB as a minimum, usually 100 – 120dB) and this can easily tempt the designer into ignoring the effect of A_{OL} entirely.

Sagging A_{OL}

A_{OL} does, in fact, change quite markedly with both frequency and temperature. We have already seen that the ac A_{OL} rolls off at a constant rate, usually 20dB/decade, and this determines the gain that can be achieved for any given bandwidth. In fact when the frequency starts approaching the maximum bandwidth the excess gain available becomes progressively lower and this affects the validity of the high-A_{OL} approximation.

As an example, take $\beta = 0.01$ (for a gain of 100) and $A_{OL} = 10^5$ at dc. The actual gain, from the feedback equation, is

$$A_{CL} = 10^5/(1 + 10^5 \cdot 0.01) \qquad = 99.9$$

Now raise the frequency to the point at which it is a decade below the maximum expected bandwidth at this gain. This will have reduced A_{OL} to ten times the closed-loop gain or 1000. The actual gain is now

$$A_{CL} = 1000/(1 + 1000 \cdot 0.01) \qquad = 90.9$$

which shows a ten-percent gain reduction at one-tenth the desired bandwidth!

A_{OL} also changes with temperature. The data sheet will not always tell you how much, but it is common for it to halve when going from the low temperature extreme to the high extreme. If your circuit is sensitive to changes in closed-loop gain, it would be wise to check whether the likely changes it will experience in A_{OL} are acceptable and if not, either reduce the closed-loop gain to give more gain margin, or find an op-amp with a higher value for A_{OL}.

5.2.13   Noise

A perfect amplifier with perfect components would be capable of amplifying an infinitely small signal to, say, 10V p-p with perfect resolution. The imperfection which prevents it from doing so is called noise. The noise contribution of the amplifier circuit places a lower limit on the resolution of the desired signal, and you will need to account for it when working with low-level (sub-millivolt) signals or when the signal-to-noise ratio needs to be high, as in precision amplifiers and audio or video circuits.

There are three noise sources which you need to consider:

• amplifier-generated noise

- thermal noise
- extraneous noise

The third of these is either electromagnetically coupled into the circuit conductors at rf, or by common-mode mechanisms at lower frequencies, or via other more esoteric routes. It can be minimised by good layout and shielding and by keeping the operating bandwidth low, and is mentioned here only to warn you to keep it in mind when thinking about noise.

Thermal noise

The other two sources, like the dc offset and bias error components discussed earlier, are conventionally referred to the op-amp input. Thermal, or "Johnson" noise is generated in the resistive component of any circuit impedance by thermal agitation of the electrons. All resistors around the input circuit contribute this. It is given by

$$e_n = \sqrt{(4kTBR)}$$

where
e_n = rms value of noise voltage
k = Boltzmann's constant, 1.38×10^{-23} joules/°K
T = absolute temperature
B = bandwidth in which the noise is measured
R = circuit resistance

As a rule of thumb, it is easier to remember that the noise contribution of a $1k\Omega$ resistor at room temperature (298°K) in a 1Hz bandwidth is 4nV rms. The noise is proportional to the square root of bandwidth and resistance, so a $100k\Omega$ resistor in 1Hz, or a $1k\Omega$ resistor in 100Hz, will generate 40nV. Noise is a statistical process. To convert the rms noise to peak-to-peak, multiply by 6.6 for a probability of less than 0.1% that a peak will exceed the calculated limit, or 5 for a probability of less than 1%.

Amplifier noise

Amplifier noise is what you will find specified in the data sheet (sometimes; where it is not specified it can be 2 – 4 times worse than an equivalent low-noise part). It is characterised as a voltage source in series with one input, and a current source in parallel with each input, with the amplifier itself being considered noiseless. The values are specified at unity bandwidth, as rms nanovolts or nanoamps per root-Hertz; alternatively they may be specified over a given bandwidth. Because you need to add together all noise contributions, it is usually easiest to calculate them at unity bandwidth and then multiply the overall result by the square root of the bandwidth. This assumes a constant noise spectral density over the bandwidth of interest, which is true for resistors but may not be for the op-amp (see later). Noise, being statistical, is added on a root-mean-square basis. So the general noise model for an op-amp circuit is as shown in Figure 5.15.

When the noise is added in rms fashion, if any noise source is less than a third of another it can be neglected with an error of less than 5%. This is a useful feature to remember with complex circuits where it is difficult to account accurately for all generator resistances.

As an example of how to apply the noise model, let us examine the tradeoffs between a high-impedance and a low-impedance circuit for different op-amps. The circuit is the standard inverting configuration with R1 sized according to the principle laid out earlier for minimization of bias current errors. R_{IN} is the sum of generator output impedance and amplifier input resistor. The op-amps chosen have the following noise characteristics (at

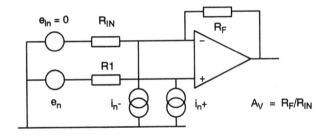

	Cause	Output voltage contribution	
R_{IN}	Thermal noise	$\sqrt{(4kTR_{IN})} \cdot A_V \cdot \sqrt{B}$	$= N(R_{IN})$
$R1$	Thermal Noise	$\sqrt{(4kTR1)} \cdot (A_V + 1) \cdot \sqrt{B}$	$= N(R1)$
R_F	Thermal Noise	$\sqrt{(4kTR_F)} \cdot \sqrt{B}$	$= N(R_F)$
i_n-	Amplifier Current Noise	$i_n- \cdot R_F \cdot \sqrt{B}$	$= N(i_n-)$
i_n+	Amplifier Current Noise	$i_n+ \cdot R1 \cdot (A_V + 1) \cdot \sqrt{B}$	$= N(i_n+)$
e_n	Amplifier Voltage Noise	$e_n \cdot (A_V + 1) \cdot \sqrt{B}$	$= N(e_n)$

Total output noise $= \sqrt{[N(R_{IN})^2 + N(R1)^2 + N(R_F)^2 + N(i_n-)^2 + N(i_n+)^2 + N(e_n)^2]}$

Figure 5.15 The op-amp noise model

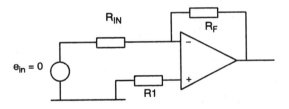

1kHz):

OP27: $e_n = 3nV/\sqrt{Hz}$ $i_n = 0.4pA/\sqrt{Hz}$ (very low noise precision bipolar)

TL071: $e_n = 18nV/\sqrt{Hz}$ $i_n = 0.01pA/\sqrt{Hz}$ (low noise biFET)

LM301A: $e_n = 16nV/\sqrt{Hz}$ $i_n = 0.27pA/\sqrt{Hz}$ (industry standard bipolar)

Working from the noise model of Figure 5.15, the contributions (in $nV/\sqrt{Hz}$) are tabulated for a low-impedance circuit and a high-impedance circuit, with the major contributor in each case shown emphasized and the negligible contributors shown in brackets:

Low impedance, $R_{IN} = 200\Omega$, $R1 = 180\Omega$, $R_F = 2K\Omega$

Noise contributor	OP27	TL071	LM301A
$N(R_{IN})$	17.9	17.9	17.9
$N(R1)$	18.7	18.7	18.7
$N(R_F)$	*(5.6)*	*(5.6)*	*(5.6)*
$N(i_n-)$	*(0.8)*	*(0.02)*	*(0.54)*
$N(i_n+)$	*(0.79)*	*(0.02)*	*(0.53)*
$N(e_n)$	**33**	**198**	**176**
Total noise voltage	41.9	200	178

High impedance, R_{IN} = 200KΩ, R1 = 180KΩ, R_F = 2MΩ

Noise contributor	OP27	TL071	LM301A
N(R_{IN})	565	565	565
N(R1)	590	**590**	**590**
N(R_F)	178	178	178
N(i_n–)	**800**	*(20)*	540
N(i_n+)	792	*(19.8)*	534
N(e_n)	*(33)*	198	176
Total noise voltage	1402	836	1143

Some further rules of thumb follow from this example:

- high impedance circuits are noisy
- in low impedance circuits, op-amp voltage noise will be the dominant factor
- in high-impedance circuits, one or other of resistor noise or op-amp current noise will dominate: use a biFET and delete R1
- don't expect an ultra-low-noise-voltage op-amp to give you any advantage in a high impedance circuit

Noise bandwidth

Deciding the actual noise bandwidth is not always simple. The bandwidth used in the noise calculations is a notional "brick-wall" value which assumes infinite attenuation above the cut-off frequency. This of course is not achievable in practice, and the circuit bandwidth has to be adjusted to reflect this fact. For a single-pole response with a cut-off frequency f_c and a roll-off of 6dB/octave, the noise bandwidth is $1.57f_c$. For a cascade of single-pole filters the ratio of the noise bandwidth to cut-off frequency decreases.

For more complex circuits it is usually enough to make some approximation to the actual bandwidth. If the low-frequency cut-off is more than a decade below the high-frequency one then it can be neglected with little error, and the noise bandwidth can be taken as from dc to the high-frequency cut-off. The exception to this is in very-low-frequency and dc applications (below a few tens of Hz), because at some point the op-amp noise contribution starts to rise with decreasing frequency. This region is known as 1/f or "flicker" noise. All op-amps show this characteristic, but the point at which the noise starts to rise (the 1/f noise corner) can be reduced from a few hundred Hz to below 10Hz by careful design of the device.

5.2.14   Supply current

Circuit diagrams often leave out the supply connnections to op-amp packages, for the very good reason that they create extra clutter, and the purpose of a circuit diagram is to communicate information as clearly as it can. When a single supply or a dual-rail supply is used throughout a circuit then confusion is unlikely, but with several different voltage levels in use it becomes difficult to work out exactly which op-amp is supplied by what voltage, and it is then better practice to show supplies to each package.

One of the dis-benefits of not showing supply connections is that it is easy to forget about supply current (I_S). Data sheets will normally give typical and maximum figures for I_S at a specified voltage, and no load. If the supply voltages are the same in the circuit as on the data sheet, and if none of the outputs are required to deliver any significant current, then it is reasonable just to add the maximum figures for all the devices in circuit to arrive at a worst-case power consumption. At other supply voltages you will have to make some estimate of the true supply current, and some data sheets include a graph of typical I_S versus supply voltage to aid in this. Also, note that I_S varies with temperature, usually increasing with cold.

When an op-amp output drives a load, be it resistive, capacitive or inductive, the current needed to do so is drawn from one supply rail or the other, depending on the polarity of the output. In the worst case of a short circuit load, I_S is limited by the device's output current limiting. It is quite possible for the load currents to dominate power supply drain. With typical quiescent I_S figures of a milliamp or so, you only need an output load resistance of $10K\Omega$ being driven with a $\pm10V$ swing to double the actual current consumption of the circuit. When calculating worst-case load currents in these circumstances, you need to know not only the maximum output swing into resistive loads, but also the current that may be needed to drive capacitive loads.

I_S vs. speed and dissipation

Op-amp supply current is usually a trade-off against speed. You can find devices which are spec'd at $10\mu A$ I_S, but such a part can only offer a slew rate of $0.03V/\mu s$. Conversely, fast devices require more current, often up to 10mA. At these levels, package dissipation rears its head. An op-amp run at $\pm15V$ with 10mA I_S is dissipating 300mW. With a thermal resistance of $100 - 150°C/W$ (the data sheet will give you the exact value) its junction temperature will be $30 - 45°C$ above ambient (see section 9.5.1), and this is before it drives any load! This could well prevent the use of the part at high ambient temperatures, and will also affect other parameters which are temperature sensitive. With such a device, make sure you know what its operating temperature will be before getting deeply involved in performance calculations.

5.2.15   Temperature ratings

And so we come naturally to the question of over what temperature range can you use a particular device. Analogue ICs historically have been marketed for three distinct sectors, with three specified temperature ranges:

- Commercial: 0 to +70°C
- Industrial: −40 to +85°C (occasionally −25 to +85°C)
- Military: −55 to +125°C

The picture is nowadays slightly blurred with the introduction of parts for the automotive market, which may be spec'd over −40 to +125°C, and with some Japanese suppliers (predominantly in the digital rather than analogue area) offering non-standard ratings such as −20 to +75°C.

If you are designing equipment for the typical commercial environment of 0 to 50°C then you are not going to worry much about device temperature ratings: just about every IC ever made will operate within this range. At the other extreme, if you are designing for military use then you will be buying military qualified components, paying the earth for them and this book will be of little use to you. But the question quite frequently arises, what parts should I use when my ambient temperature range goes a few degrees below zero or above 70°C?

In theory, you should use industrial-temperature-rated devices. Unhappily, there are three good reasons why you might not:

- the part you want to use may not be available in the industrial range;
- if it is available, it may be too expensive;
- even if it is listed as available, it may actually prove to be on a long lead-time or otherwise hard to get.

So the question resolves itself into: can I use commercial parts outside their specified temperature range? And the answer is: maybe. No IC manufacturer will give you a guarantee that the part will operate outside the temperature range that he specifies. But the fact is, most parts will, and there are two main factors which limit such use, namely specifications and reliability.

Specification validity

The manufacturer will specify temperature-sensitive parameters (which is most of them) either at a nominal temperature (25°C) or over the temperature range. These specifications have bite, in that if the part fails to meet them the customer is entitled to return it and ask for a replacement. So the manufacturer will test the parts at the specification limits. However, he is not responsible for what happens outside the temperature range, and it is more than likely that some parameters will drift out of their specification when the temperature limits are over-stepped. Very often these parameters are unimportant in the application, such as offset voltage in an ac amplifier. Therefore you can with care design a circuit with wider tolerances than would be needed for the published figures and trust that these will be sufficient for wide-temperature range abuse.

It is of course a risky approach, but it may be necessary in some circumstances. Two extra risks are that some parameters may change much more outside the specified temperature range than they do within it, and that you may successfully test a sample of manufacturer A's product, but manufacturer B's nominally identical parts behave quite differently. We shall comment on this again in section 5.2.16.

Package reliability

The second factor is reliability. The reliability of any semiconductor device worsens with increasing temperature; a temperature rise of 10°C halves the expected lifetime. So operating ICs at high temperatures is to be avoided wherever possible, but there is no magic cut-off at 70°C or 85°C. The maximum junction temperature should always be observed, but this is usually in the region of 100 – 150°C.

At low temperatures the problem is included moisture. Moulded plastic packages allow some moisture to creep along the lead-to-plastic interface (this is worse at high temperatures and humidities) and this can accumulate over the surface of the chip, where it is a long-term corrosive influence. When the operating temperature dips below 0°C the moisture freezes, and the resulting change in conductivity and volume can give sudden changes in parameters which are well outside the drift specifications. The effect is very much less with "hermetic" packages using a glass-ceramic-metal seal, and in fact progress in plastic packages has advanced to the point where included moisture is not as serious a problem as it used to be. Other board-related problems arise when equipment is used below 0°C due to condensation of airborne moisture on the cold pcb surface, as ambient temperature rises.

5.2.16   Cost and availability

The subtitle to this section could be, why use industry standards? Basically, the

Conventional bipolars

	LM2902	LM301A	LM308	LM318	LM324	LM348	LM358	MC3403	MC1458	RC4136	RC4558	NE5532	NE5534	709	741	747	748	776
Motorola	■						■	■	■					■	■	■	■	■
Signetics		■	■		■	■	■								■	■		
Texas	■	■			■		■	■	■		■		■		■	■	■	
STM	■	■	■	■	■	■	■	■	■	■	■	■	■	■	■		■	■
National	■	■	■	■	■	■	■	■	■		■	■		■	■		■	■
Raytheon	■	■	■	■	■	■			■	■	■	■	■	■	■		■	
GE/RCA	■	■	■		■		■	■	■	■				■	■	■	■	
Lin Tech			■	■	■													
Avge cost	0.29	0.25	0.46	1.20	0.24	0.28	0.20	0.47	0.19	0.45	0.28	0.52	0.44	0.46	0.19	0.38	0.30	0.53
No of sources	6	6	5	4	7	4	6	5	6	3	4	3	3	4	7	6	5	3

BiFETs

	LF347	LF351	LF353	LF355	LF411	LF412	LF441	LF442	TL061	TL062	TL064	TL071	TL072	TL074	TL081	TL082	TL084
Motorola	■														■	■	■
Texas		■	■		■	■	■	■		■	■	■	■	■	■	■	■
STM		■	■	■					■	■	■	■	■	■	■	■	■
National	■	■	■	■	■	■	■	■	■	■	■	■	■	■	■	■	
Avge cost	0.37	0.25	0.28	0.70	0.55	0.95	0.42	0.64	0.32	0.34	0.47	0.29	0.32	0.46	0.27	0.31	0.46
No of sources	2	3	3	2	2	2	2	2	2	3	3	3	3	3	4	4	3

Table 5.1 Industry standard op-amps

Precision & Chopper

	OP-07D	OP-27G	OP-37G	ICL7650	ICL7652
PMI	■	■	■		
Analog Dev	■	■	■		
Texas	■	■	■		
Raytheon	■	■	■		
Lin Tech	■	■	■		■
Motorola		■	■		
GE/Intersil				■	■
Maxim	■			■	■
Teledyne				■	■
Siliconix					■

Table 5.2   Industry standard precision and chopper op-amps

application of op-amps (along with virtually all other components) follows the 80/20 rule beloved of management consultants: 80% of applications can be met with 20% of the available types. These devices, because of their popularity, become "industry standards" and are sourced by several manufacturers. Their costs are low and their availability is high. The majority of other parts are too specialised to fulfil more than a handful of applications and they are only produced by one or perhaps two manufacturers. Because they are only made in small quantities their cost is high, and they can sometimes be out of stock for months.

As an indication of what may be regarded as "industry standard", Table 5.1 lists a range of devices along with their manufacturers and an average 100-up price for each. The criterion for inclusion in this table was that each device should be offered by three or more manufacturers. Similarly, Table 5.2 lists some industry-standard "precision" op-amps.[†]

When to use industry standards

The virtue of selecting from these tables is that the parts are well established, unlikely in the extreme to run into sourcing problems or be withdrawn (the humble 741 has been around for over 20 years!) and, because of the competition between manufacturers, they will remain cheap. If they will do the job, use them in preference to a sole sourced device. But, nothing comes for free: the negative aspect of multi-sourcing is that many parameters go unspecified for cheap devices, and this leaves open the possibility that different manufacturers' nominally identical parts can differ substantially in those parameters that are omitted from the common spec. As an example, try to find a slew-rate specification for the LM324: you won't. If you've designed and tested a circuit with manufacturer A's devices, and they happen to be quite fast, you will be heading for production problems when your purchasing manager buys a few thousand of manufacturer B's devices which are slower.

There are two approaches here. Either, design the circuit from the outset to be

---

†. Omission of a device from these tables is not meant to imply that it *will* be expensive and hard-to-get. Some manufacturers have a good reputation for reliability of supply and pricing, even of sole-sourced items.

insensitive to those parameters which are badly specified, un-specified or (worse) specified differently in the data sheets of each manufacturer. Or, test all the different manufacturers' products and allow purchasing to buy only those which live up to your expectations. Needless to say, the first course is infinitely preferable.

Quad or dual packages

Comparing prices in the tables, you can see that the LM324 does offer, in fact, the lowest cost-per-op-amp (5p). This points up another factor to bear in mind when selecting devices: choose a quad or dual package in preference to a single device, when your circuit uses several gain stages. This reduces both unit cost and production cost. The disadvantages are inflexibility in supply voltage and pc layout, and possible thermal, power rail or rf interaction between gain blocks on a single substrate. It is rare for any of these to be serious enough to offset the advantages.

5.3 Comparators

A comparator is just an op-amp with a faster slew rate, and with its output optimised for switching. It is intended to be used open-loop, so that feedback stability considerations don't apply. The device exploits the very large open-loop gain of the op-amp circuit so that the output swings between "fully-on" and "fully-off", depending on the polarity of the differential input voltage, and there should be no stable state in between. Input-referred and open-loop parameters – offsets, bias currents, temperature drift, noise, common-mode and power-supply rejection ratio, supply current and open-loop gain – are all specified in the same way as op-amps. Output and ac parameters are specified differently.

5.3.1   Output parameters

The most frequent use of a comparator is to interface with logic circuitry, so the output circuit is designed to facilitate this. Two configurations are common: the open collector, and the totem pole (Figure 5.15). The open-collector type requires a pull-up

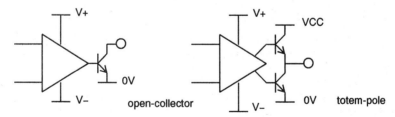

Figure 5.15 Comparator outputs

resistor externally, while the totem-pole does not. Both types interface easily to the classical LSTTL logic input, which requires a higher pull-down current than is needed to pull it up. The output is specified either in terms of its saturation voltage, sink current, leakage current and maximum collector voltage for the open-collector type or in terms of high- and low-level output voltages at specified load currents for the totem-pole type.

Because the totem-pole type is invariably aimed at logic applications, it is always specified for 5V output levels. The open-collector type, which includes the highly popular LM339 and LM311, is far more flexible since any output voltage can be obtained simply by pulling up to the required rail, which can be separate from the analogue supply rails.

5.3.2   AC parameters

Because the comparator is used as a switch, the only ac parameter which is specified is the response time. This is the time between an input step function and the point at which the output crosses a defined threshold. It includes the propagation delay through the IC and the slewing rate of the output. Outside of the device itself, two factors have a large effect on the response time:

- the input overdrive
- the output load impedance

Overdrive

For the specifications, an input step function is applied which forces the differential input voltage from one polarity to the other. The overdrive, as in Figure 5.17, is the final steady-state differential voltage. Usually, the step amplitude is held constant and its

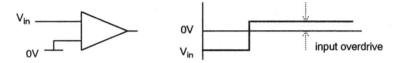

Figure 5.17 Comparator overdrive

offset is varied to give different overdrive values. The greater the overdrive, the more current is available from the differential input stage to propagate the change of state through to the output, although beyond a certain point there is no gain to be had from increasing it. Small overdrives can lead to suprisingly long response times and you should check the data sheet carefully to see if the device is being specified in similar fashion to how your circuit will drive it.

The specification test assumes that the step function has a much shorter rise-time than the response to be measured. Response time specs are virtually meaningless when the comparator is driven by slow rise-time analogue signals. We shall discuss this more fully under the heading of hysteresis.

Load impedance

The output load resistance R_L (for open-collector types) and capacitance C_L have a major influence on the output slewing rate. The capacitance includes the device output capacitance, circuit strays and the input capacitance of the driven circuit (this last is usually the most significant). The slewing rate is determined by the current that is available to charge and discharge the capacitance, following the rule $dV/dt = I/C$. For the negative-going transition this current is supplied by the output sink transistor and is in the region of $10 - 50mA$, assuring a fast edge, but the current available to charge the positive transition is supplied by the pull-up device or resistor and may be an order of magnitude lower. The choice of output resistor directly affects the positive-going rise time (Figure 5.18) and the power dissipation of the circuit.

The advantages of the active low

On this latter point, it is worth remembering that if you expect low-duty-cycle pulses at

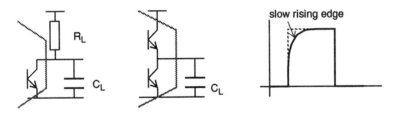

Figure 5.18 Output slewing vs. load capacitance

the output, want low power drain and a fast leading edge and have a choice of logic polarity, that the preferable configuration is to use an active-low output as in Figure 5.19(a). The signal is normally off so that power drain is low, and the leading edge transition depends on the output transistor rather than the pull-up. If a fast trailing edge is also needed, the pull-up can be reduced in value without significantly affecting power drain if the duty cycle is low. It is easy and cheap to provide a logic inverter if you really need positive going pulses.

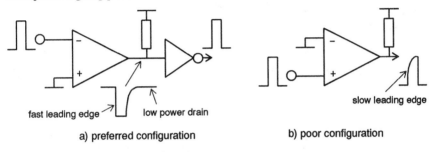

Figure 5.19 Comparator output configurations

Pulse timing error

Continuing this train of thought, you can see that it is quite easy for the pulse timing to be affected by the output rise- and fall-times. This is quite often the source of unexpected errors in circuits which convert analogue levels into pulse widths for timing measurement. Because the pulse rising edge is slowed to a greater extent than the falling edge, the point at which it crosses the following logic gate's switching threshold is different, so that rising and falling analogue inputs result in different switching points. This effect is demonstrated in Figure 5.20. The problem is generally more

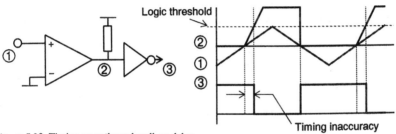

Figure 5.20 Timing error through pull-up delay

visible with CMOS-input-level gates than it is with TTL-input-level ones, as TTL's switching threshold is closer to 0V whereas the CMOS threshold is ill-defined, being anywhere between 0.3 and 0.7 times its supply rail. The difference can amount to a microsecond or more in low power circuits.

5.3.3   Op-amps as comparators (and vice versa)

You may often be faced with a circuit full of multiple op-amp packages and the need for a single comparator. Rather than invest in an extra package for the comparator function, it is quite in order to use a spare op-amp as a comparator with the following provisos:

- the response time and output slew-rate are adequate. Typical standard op-amp slew rates of $0.5V/\mu s$ will traverse the logic "grey area" from 0.8 to 2V in about $3\mu s$; this is too slow for some logic functions. Uncompensated op-amps without compensation capacitors make much better comparators.

- in some op-amps, recovery from the saturated state can take some time, causing appreciable delays before the output starts to slew. This is hardly ever specified on data sheets.

- the output voltage swing and drive current are adequate and correct for the intended load. Clearly you cannot drive 5V logic directly from an op-amp output that swings to within 2V of ±15V supply rails. Some form of interface clamping is needed; this could take the form of a feedback zener arrangement so that the output is not allowed to saturate, which confers the additional benefit of reduced response time. Also, it is not good practice to drive TTL-level inputs directly from a LM324-style output even from a single 5V analogue supply without some additional buffering, as these outputs cannot sink TTL input currents and still remain safely within the TTL low voltage threshold of 0.8V. They are perfectly happy with CMOS inputs, though.

It is also possible, if you have to, to use a comparator as an op-amp. (In most cases: some totem-pole outputs cannot be operated in the linear mode without drawing destructively large supply currents.) It was never designed for this, and will be hideously unstable unless you slug the feedback circuitry with large capacitors, in which case it will be slow. Also, of course, it is not characterised for the purpose, so for some parameters you are dealing with an unknown quantity. Unless the application is completely non-critical it is best to design op-amp circuits with op-amps.

5.3.4   Hysteresis and oscillations

When the analogue input signal is changing relatively slowly, the comparator may spend appreciable time in the linear mode while the output swings from one saturation point to the other. This is dangerous. As the input crosses the linear-gain region the device suddenly becomes a very-high-gain open-loop amplifier. Only a small fraction of stray positive feedback is needed for the open-loop amplifier to become a high-frequency oscillator (Figure 5.21).

The frequency of oscillation is determined by the phase shift introduced by the stray feedback and is generally of the same order as the (uncompensated) equivalent unity gain bandwidth. This is not specified for comparators, but for typical industry-standard devices is several of MHz. The term "relatively slowly" as used above means relative

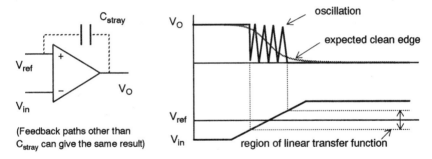

Figure 5.21 Oscillation during output transitions

to the period of the oscillation, so that any traverse of the linear region which takes longer than a few hundred nanoseconds must be regarded as slow: this of course applies to a very large proportion of analogue input signals!

The subtle effects of edge oscillation

This oscillation can be particularly troublesome if you are interfacing to fast logic circuits, especially when connecting to a clock input. It can be hard to spot on the scope, as you will probably have the timebase set low for the analogue signal frequency, but the oscillations appear to the digital input as multiple edges and are treated as such: so for instance a clock counter might advance several counts when it appears to have had only one edge, or a positive-going clock input might erroneously trigger on a negative-going edge.

Even when you don't have to contend with high-speed logic circuits, the oscillation generated by the comparator can be an unexpected and unwelcome source of rf interference.

Minimise stray feedback

The preferred solution to this problem is to reduce the stray feedback path to a minimum so that the comparator remains stable even when crossing the linear region. This is achieved by following three golden rules:

- keep the input drive impedance low;
- minimise stray feedback capacitance by careful layout;
- avoid introducing other spurious feedback paths, again by careful layout and grounding.

The lower the input impedance, the more feedback capacitance is needed to generate enough phase shift for instability. For instance, 2pF and 10KΩ gives a pole frequency of 8MHz, a perfectly respectable oscillation frequency for many high-speed comparators. It is hard to reduce stray capacitance much below 2pF, so the moral is, keep the drive impedance below 10KΩ, and preferably an order of magnitude lower.

Minimum stray capacitance from output to input should always be the layout designer's aim; follow the rules quoted in section 5.2.11 for high-frequency op-amp stability. Most IC packages help you in this regard by not putting the output pin close to the non-inverting input pin. Don't look this particular gift horse in the mouth by running the output track straight back past the inputs! Guarding the inputs (see section 2.3.1) can be useful. And, again as with op-amp circuits, do not introduce ground-loop or common-mode feedback paths by incorrect layout.

Hysteresis

Another approach to the problem of unwanted oscillation is to kill it with hysteresis. This approach is used when the above methods fail or cannot be applied, and you can also use it as a legitimate circuit technique in its own right, as in the well-known Schmitt trigger. Hysteresis is the application of deliberate positive feedback in order to propel the output speedily and predictably through the linear region. The principle of hysteresis is shown in Figure 5.22.

Note that although this looks superficially like the classic inverting op-amp configuration, feedback is applied to the *non*-inverting input and is therefore operating in the positive sense. Note also that the application of hysteresis modifies the switching threshold in both directions, and that it is modified differently in either direction by the presence of R3. This resistor is shown in the circuit of Figure 5.22 to emphasize that it must be included in calculating hysteresis; we have assumed that the comparator is the open-collector type. If the output is the totem-pole type, then R3 is omitted but the output levels and impedance must be taken into consideration. These values directly affect the switching threshold and can cause surprisingly large inaccuracies.

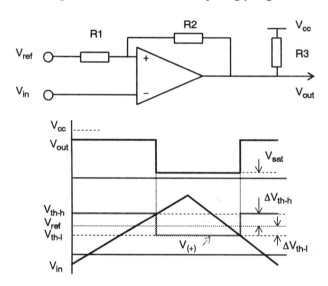

Ignoring input and output leakage currents,

$$V_{out} (H) = V_{cc} - (V_{cc} - V_{ref})(R3/[R1 + R2 + R3])$$
$$V_{out} (L) = V_{sat}$$

$$V_{th\text{-}h} = \alpha \cdot V_{cc} + (1 + \alpha) \cdot V_{ref} \qquad \text{where } \alpha = R1/(R1 + R2 + R3)$$
$$\Delta V_{th\text{-}h} = \alpha \cdot (V_{cc} + V_{ref})$$

$$V_{th\text{-}l} = \beta \cdot V_{sat} + (1 - \beta) \cdot V_{ref} \qquad \text{where } \beta = R1/(R1 + R2)$$
$$\Delta V_{th\text{-}l} = \beta \cdot (V_{sat} - V_{ref})$$

A common simplification is that R3 << R1+R2 so that $\alpha \approx \beta$, and that V_{ref} is half of V_{cc} and $V_{sat} = 0$, in which case ΔV_{th} (the total hysteresis band) reduces to $\beta \cdot V_{cc}$.

Figure 5.22 Hysteresis

Because hysteresis deliberately alters the switching threshold, it cannot be indiscriminately applied to all comparator circuits to clean up their oscillatory tendencies, nor should it. The techniques outlined previously should be the first priority. But it is not always possible to keep drive impedances low and where high impedance is necessary, hysteresis is a valuable tool. If the minimum input dV/dt is predictable, you can also apply a judicious amount of AC hysteresis (by substituting a capacitor for R2) which will prevent oscillation without affecting the DC threshold: but beware slow-moving inputs or you will simply end up with a longer time-constant oscillator!

5.3.5   Input voltage limits

When an op-amp is operating closed-loop, the differential voltage at its inputs is theoretically zero. If it isn't then the feedback loop is open, either by design or because of one or another form of overloading. Comparators on the other hand are intended for open-loop operation and their differential input voltage is never expected to be zero.

Data sheets specify the maximum voltage range of differential input signals and this should not be overlooked. If it is exceeded, too much current through the breakdown of the input transistor base-emitter junctions (or MOS gates) can degrade the input offset and bias current parameters. Most of the industry-standard LM339 derivatives have a differential limit equal to the supply rail limit, but some comparators have quite restricted differential input ranges. For instance the fast NE529 has a differential input restriction of ±5V, with a common mode of ±6V. These two quantities interact: both inputs at +4V will satisfy the common-mode limit, but if one is left at +4V the other cannot be taken below −1V because the differential voltage is then greater than 5V.

Even if the normal operating differential range is kept within limits, it is possible for abnormal conditions (such as cycling of separate power rails) to breach the limit. If this is at all likely, and if the condition can't be prevented, at the very least include some input current protection resistance. You can calculate the required values from the expected or possible over-voltage divided by the absolute maximum input current, or from the power dissipation, which is always quoted on device data sheets.

Comparator parameters vs. input voltage

Also, while considering large differential input voltages, remember that unexpected things can happen to the comparator even when the limits are not exceeded. Response time is usually specified for a common-mode voltage of zero and may degrade when the common-mode limits are approached; this applies equally to bias currents. Some data sheets show curves of input bias current which have step changes (Figure 5.23) at certain differential input voltages, due to internal dc feedback. Notice these and make sure your circuit can cope!

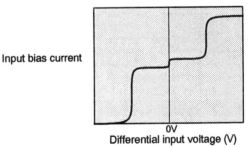

Input bias current

0V

Figure 5.23 Input bias current steps   Differential input voltage (V)

Comparators	LM311	LM319	LM339	LM393	LM2903	MC3302
National	■	■	■	■	■	■
Motorola	■		■	■	■	■
Signetics	■	■	■	■	■	■
GE/RCA	■		■			■
Texas	■		■	■	■	■
STM	■	■	■	■	■	■
Raytheon	■		■	■		■
Lin Tech		■				
Average cost	0.22	0.97	0.23	0.23	0.31	0.43
No. of sources	7	4	7	6	5	7

Table 5.3 Industry standard comparators

In multi-comparator packages, some comparators may remain unused. Never leave unused inputs open, as that device could oscillate on its own, which would then be coupled into the other devices in the same package. If both inputs are grounded, the unpredictable offset voltage will mean that the output voltage, and hence unit supply current, will vary. The safest course is to ground one input and supply the other from another fixed voltage within its differential and common mode limit (which might include the supply rail), so that the device is always saturated.

5.3.6   Comparator sourcing

Exactly the same comments about sourcing apply to comparators as were made earlier about op-amps (see section 5.2.16). Table 5.3 is included here to show the most common comparator types available. Like the LM324 op-amp, the most popular and cheapest part per comparator is the quad LM339.

5.4   Voltage references

The need for a stable reference voltage is found in power supplies, measurement instrumentation, DAC/ADC systems and calibration standards. Two techniques exist to provide such references, one based on the precision zener diode and the other on the band-gap voltage of silicon.

5.4.1   Zener references

We have already discussed the operation of the basic Zener diode (section 4.1.7). To produce a reference from a Zener, it must be temperature compensated, fed from a constant current and buffered. Temperature compensation is achieved by selecting a low-tempco zener voltage in the range 5.5 – 7V and mating it with a silicon diode so that the voltage tempcos cancel. The combination is driven from a constant current generator and buffered to give a constant output voltage regardless of load.

Since surface breakdown increases noise and degrades stability, a precision zener is usually fabricated below the surface of the IC which contains its support circuitry, but this gives a greater spread of tempco and absolute voltage. The overall reference must therefore allow adjustment of these parameters, normally by laser wafer trimming. Such references can offer long-term stability of 50ppm/year and absolute accuracy of 0.1% with ±10ppm/°C tempco. Better performance is obtained if the reference can be stabilised with an on-chip heater, as in the LM399 for example. This takes a comparatively large power drain and has a warm-up time measured in seconds but offers sub-ppm tempcos.

5.4.2   Band-gap references

A significant disadvantage of the zener reference is that its output voltage is set at around 6.9V and it therefore needs a comparatively high supply voltage. A competing type of reference overcomes this and other problems, notably cost and supply current, and has become extremely widespread since its invention by Robert Widlar in 1971. The fundamental circuit is shown in Figure 5.24. In this circuit I1 and I2 differ by a fixed ratio and V_{ref} is given (neglecting base currents) by

$$V_{ref} = V_{BE3} + I2 \cdot R2 = V_{BE3} + (V_{BE1} - V_{BE2}) \cdot R2/R1$$

The temperature coefficient of the second term can be arranged by suitable selection of I1, R1 and R2 to cancel that of the V_{BE3} term. This turns out to occur when V_{ref} is in the neighbourhood of 1.2V, which is equivalent to the "band-gap" voltage of a silicon junction at 0°K.

Such a band-gap reference, relying only on matched transistors, is easily integrated along with biasing, buffer and amplifier circuitry to give a complete reference in a single package. It is capable of a lower minimum operating current and a sharper knee than any zener. As well as the un-processed band-gap voltage of 1.2V (actual voltage depends on detailed internal design and process variations and varies between 1.205 and 1.26V) devices are available with trimmed outputs of 2.5, 5 and 10V, principally for use in digital-to-analogue/analogue-to-digital conversion circuits.

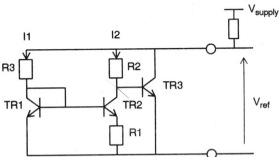

Figure 5.24 The band-gap reference

Costs and interchangeability

There is an obvious trade-off between initial voltage tolerance and tempco on the one hand, and cost and availability on the other, since the manufacturer has to accept a lower yield and longer test and trim time for the closer tolerances. Initial voltage can be trimmed exactly with a potentiometer, but this method adds both parts and production cost which will offset the higher cost of a tighter-tolerance part. Trimming the reference

Type	Output voltage	Tolerance	Tempco	Min. current	Cost £, 25+
MP5010GN	1.22V	2.5%	30ppm/°C	50μA	0.65
ICL8069DCZR	1.23V	1.6%	100ppm/°C	50μA	0.88
ICL8069CCZR	1.23V	1.6%	50ppm/°C	50μA	1.22
LM385Z-1.2	1.235V	2%	20ppm/°C avg	10μA	1.39
LM385BZ-1.2	1.235V	1%	20ppm/°C avg	10μA	2.33
TSC9491BJ	1.22V	2%	30ppm/°C typ	50μA	0.84
TSC04BJ	1.26V	1.6%	30ppm/°C typ	15μA	0.89
REF12Z	1.26V	1%	40ppm/°C typ	90μA	0.55

Table 5.4 Two-terminal 1.2V voltage references

voltage can also worsen the reference temperature coefficient in some configurations, and there is the extra tempco of the trimming components to include. Table 5.4 shows a sample of typical two-terminal 1.2V references, including their tolerance, tempco, minimum operating current and cost.

Although it would appear from this table that there is a wide choice of types offering much the same performance, not all of these are directly interchangeable. The minor differences in regulation voltage may catch you out if you have designed a circuit for a given voltage tolerance and subsequently want to change to a different type. The preferable solution is to allow as wide a tolerance as possible in the first place. Also, there are variations in the allowable or required capacitive loading. Some parts require a decoupling capacitor of 0.1 – 1μF across them, others require that such a capacitor is *not* included. Although all those listed above are supplied in the TO-92 package and many are available in small outline, not all pin-outs are the same. Again, check before specifying alternatives.

5.4.3   Reference specifications

Line and load regulation

Line regulation is the change in output voltage due to a specified change in input voltage, normally quoted in microvolts per volt. Load regulation is a similar change due to a change in load current, expressed either in percent for a given current change or as a dynamic resistance in ohms. It should, but doesn't always, include self-heating effects due to dissipation change.

Output voltage tolerance

This is the deviation from nominal output voltage. It is quoted at a given temperature and input voltage or current, and the nominal voltage will differ under other conditions. Generally it is expressed as a percentage figure, but the asymmetry of device yields may persuade a manufacturer to quote upper and lower bounds and the nominal figure may not be in the middle of them. In your circuit design, it is best to ignore the nominal voltage and work everything out for upper and lower limits.

Ouput voltage temperature coefficient

This is the output voltage change due to an ambient temperature difference, usually from 25°C. Because neither band-gap nor zener references exhibit a straight line voltage-temperature curve (see Figure 5.25) manufacturers choose different ways to express their tempcos, sometimes as an average value across the range in ppm/°C, sometimes as different values at a series of spot temperatures, and sometimes as a worst-case error band in mV. To properly evaluate different manufacturers' references you need to correct for these differences in specification.

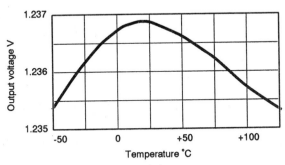

Figure 5.25 Typical band-gap reference temperature characteristics

Long-term stability

Usually expressed in ppm/1000hr or in microvolts change from the nominal voltage, this is a difficult specification to verify and so is often quoted as a typical figure based on characterisation of a sample. It is rarely specified on the cheaper components. Zeners tend to stabilise after a couple of years, so for ultra-precision applications the practice of burning in zener references at high temperatures to speed up the settling proces is sometimes followed.

Settling time

This is the time taken for the output to settle within a specified error band after application of power. It is typically in the tens to hundreds of microseconds region, and is normally only of interest if you are concerned about the dynamic performance of the reference circuit. It does not include any long-term effects due to thermal shifts, but of course these do occur, more noticeably at higher operating currents.

Minimum supply current

The regulation of a two-terminal device is not maintained below a certain minimum current. Typical values for bandgap references are 50 – 100μA, with 10μA being available although some earlier devices are much higher than this. The very low useable operating currents combined with low dynamic resistance at these currents make band-gap devices very much preferable to zener types for low-power circuitry. The maximum operating current is usually based on the point at which the device goes outside its regulation specification, but may also be determined by allowable power dissipation.

Chapter 6

Digital Circuits

The great success story of digital electronics is due to one simple fact: information can be reduced to a stream of binary data which can be represented as one of two discrete voltage levels. This data can be manipulated and processed at will, and the quantity of information you can process depends only on the speed at which you can do it. The infinite variability of analogue voltage levels is replaced by two dimensions of quantization, in voltage and time. In theory, all voltage levels below a given threshold represent binary 0, and all levels above the threshold represent binary 1. Again in theory, time is divided into discrete units by a reference clock, and the boundary between each unit marks the transition from one bit of data to the next.

By this means, the unpredictability and variability of analogue, or linear, electronic phenomena is factored out of the design process. (It is replaced by another kind of unpredictability and variability, that of complex software phenomena, but that is not the subject of this book.) Voltage drift, component tolerances, offsets and impedance inaccuracies become instantly irrelevant. At the same time, programmability allows a single piece of hardware to perform widely different tasks, including ones that perhaps were not even envisaged when it was designed and built. To incorporate such programming flexibility into analogue hardware would be impossible.

The millions of successful digital designs worldwide testify to these advantages. At the same time, much to the relief of those who would bewail the apparent soullessness of the digital universe, analogue phenomena have not been completely banished. They have merely changed their appearance. Ohm's Law still holds, and the grand laws of Electromagnetic Field Theory still maintain their grip on digital electrons, even tightening it as designers strive for ever greater speed. Variability finds its way into the gap between 1 and 0, and into the spaces between one clock period and the next. It is the digital designer's task to understand it and control it.

6.1   Logic ICs

The interfaces between logic integrated circuits including signal, clock and power supply lines must be considered to achieve a reliable digital design. This applies whether the devices concerned are microprocessors, their support chips, application-specific ICs (ASICs), programmable logic arrays (PLAs) or standard "glue" logic.

6.1.1   Noise immunity and thresholds

A logic input can take any value of voltage, nominally from one supply rail to the other, although due to transmission line effects (section 1.3) the actual voltage can exceed either supply rail on transitions. Each input is designed so that any voltage below one level, conventionally V_{IL}, is regarded as a logic "0" and any voltage above another level, V_{IH}, is regarded as logic "1" (Figure 6.1).

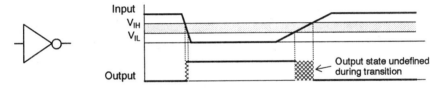

Figure 6.1 Transitions through logic thresholds

These levels are characterised for each logic or microprocessor family, and worst-case values of V_{IL} and V_{IH} can be found on any data sheet. Note that, as with any hardware-determined parameter, they may vary with temperature and you should ensure that the values you use are guaranteed across the device's temperature range. They are also a function of supply voltage. If all ICs are fed from the same supply this is not a problem, but it becomes more significant if you are interfacing logic circuits which may be fed from different supply rails.

The significance of the band between V_{IL} and V_{IH} is that the input logic state (and therefore the output state) is undefined while the voltage is in this band. Therefore transitions between logic states must happen as rapidly as possible and no decisions must be taken while the input is in transit, or for a given period (the "settling time") thereafter. This is why clocked, or synchronous, circuits are far more reliable than un-clocked or asynchronous ones for complex logic operations: the state of the clock determines when logic decisions are taken, and it is arranged that all data transitions occur when the clock is inactive.

Susceptibility to noise

Provided that all signals to logic inputs, whether from other logic outputs or from interfaces to other circuits, lie outside the V_{IL} – V_{IH} band when they are active, then in theory no misinterpretation of the input should occur. The difference between a "low" output logic level (V_{OL}) and V_{IL}, or that between a "high" output level (V_{OH}) and V_{IH}, is the noise immunity (expressed in volts) of the logic interface. Some noise immunity values of typical interfaces are given in Table 6.1. Notice that noise immunity is not a property of any particular device, but of the interface between devices. The noise immunity of a family of devices (such as LS-TTL or HCMOS) only refers to interfaces between devices of the same family.

Table 6.1 shows values of noise voltage immunity V_N for three common logic families[†], for interfaces between these families and for interfaces with a typical modern microprocessor, the Motorola MC68HC11. Since the LS-TTL family takes an appreciable dc input current in the logic low state (as do all TTL family variants), the noise immunity driving LS-TTL inputs depends on how many parallel inputs are connected. This is not the case with CMOS inputs.

Current immunity

The noise immunity value gives meaning to the ability of the interface to withstand externally-coupled noise without corruption of the perceived logic level. Thus the HCMOS-LSTTL interface can tolerate a variation of 2.4V in the high state, or 0.47V in

---

† The LS-TTL family has the generic type number 74LSXXX; HCMOS has the type number 74HCXXX, while HCTMOS has the type number 74HCTXXX. 4000B-series is self evident.

Interface		Logic high voltage immunity			Logic low voltage immunity			Current immunity (mA) = V_N/R_O		
Output to	Input	V_{OH}	V_{IH}	V_N	V_{OL}	V_{IL}	V_N	$R_O(H/L)\Omega$	High	Low
LS-TTL	10 x LS-TTL	2.7	2.0	0.7	0.4	0.8	0.4	100/25	7	16
HCMOS	HCMOS	4.4	3.15	1.25	0.1	1.35	1.25	50	25	25
4000B CMOS	4000B CMOS	4.45	3.15	1.3	0.05	1.35	1.3	800	1.625	1.625
LS-TTL	4000B/HCMOS	2.7	3.15	(-0.45)	0.4	1.35	0.95	100/25	-	38
LS-TTL	HCTMOS	2.7	2.0	0.7	0.4	0.8	0.4	100/25	7	16
HCMOS	10 x LS-TTL	4.4	2.0	2.4	0.33	0.8	0.47	50	48	9.4
4000B CMOS	1 x LS-TTL	4.4	2.0	2.4	0.4	0.8	0.4	800	3	0.5
MC68HC11	4 x LS-TTL	4.3	2.0	2.3	0.4	0.8	0.4	1000/250	2.3	1.6
MC68HC11	HCMOS	4.4	3.15	1.25	0.1	1.35	1.25	1000/250	1.25	5
LS-TTL	MC68HC11	2.7	3.15	(-0.45)	0.4	0.9	0.5	100/25	-	20
HCMOS	MC68HC11	4.4	3.15	1.25	0.1	0.9	0.8	50	25	16

Notes: all figures specified at 4.5V V_{CC} over temperature range; family characteristics taken for standard logic (not bus drivers/receivers); R_O is derived from published output characteristics

Table 6.1 Noise immunity of various logic interfaces

the low state. These are worst-case values and the actual circuit could tolerate somewhat more before a change of state occurred. But the voltage difference is only part of the story. When noise is coupled into an interface, the impedance of the interface is just as important, since this determines what voltage will be developed by a given induced interference current. The impedance is normally defined by the output driver (as long as transmission line effects are neglected) and the latter columns of Table 6.1 show the effective noise current threshold of the interface, given by the noise immunity voltage divided by the driver output impedance. This gives a truer picture of the actual noise immunity of a given combination.

Notice that the metal-gate 4000B CMOS logic family has a high output impedance at 5V compared with the other families, so that its current immunity is significantly worse. However, as the supply voltage increases so its output impedance goes down, and the combined effect means that its immunity at 15V V_{CC} is about ten times better than at 5V. It is inherently insensitive to low voltage inductively-coupled noise, but shows poor rejection of capacitively coupled noise. For general purpose 5V applications the 74HC family is preferred. You can also see that the microprocessor's high output resistance means that it does not compare favourably with standard logic. Most microprocessors are poorly specified in this respect, but the MC68HC11 is typical of modern CMOS types; older NMOS versions are no better.

Use of a pull-up

Note that the figures for high-state and low-state immunities are often different, because of the differing drive impedances and voltage thresholds in the two states. A negative immunity value indicates that, if nothing further is done, this particular interface combination will be unreliable *by design*. For instance, the 2.7V minimum high output of LS-TTL is less than the required 3.15V minimum V_{IH} for HCMOS so

LS-TTL driving directly into HCMOS is in danger of incorrectly transmitting the logic high level. The standard remedy for this particular situation is a pull-up resistor to V_{CC} to ensure a higher output from the LS-TTL (Figure 6.2). The minimum resistor value is

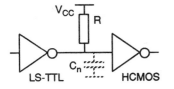

$$R_{min} = [V_{CC} - V_{OL}]/I_{OL}$$
where I_{OL} is the LS-TTL output sink current for an output voltage of V_{OL}, lower than the HCMOS low input threshold

$$R_{max} = t/C_n$$
where t is the maximum edge rise time and C_n is the node capacitance

Figure 6.2 Logic interface pull-up resistor

a function of the driver output capability, and the maximum value depends on permissible timing constraints. Alternatively, use the HCTMOS family, whose inputs are characterised especially for driving from LS-TTL levels.

Dynamic noise immunity

The static noise margins as discussed above apply until the interference approaches the operating speed of the devices. When very fast interference is present, higher amplitude is necessary to induce upset. The dynamic noise margin is measured by applying an interference pulse of known magnitude and increasing its width until the device just begins to switch. This yields a plot of noise margin versus pulse width such as shown in Figure 6.3.

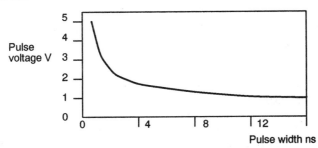

Figure 6.3 Dynamic noise immunity of 74HC series devices

You may often be forced to interface different logic families. Typically, a microprocessor may need to drive LS-TTL buffers, or you may not be able to source a particular part in the same family as the rest of the system, or you may need to change families to optimise speed/power product. You can normally expect logic interfaces of the same family to be compatible, but whenever different families or a custom interface are used you have to check the logic threshold aspects of each one.

6.1.2   Fan-out and loading

The output voltage levels that are used to fix the noise immunity thresholds of Table 6.1 are not absolute. They depend as usual on temperature, but more importantly on the output current that the driver is required to source or sink. This in turn depends on the type of loading that each output sees (Figure 6.4).

Any driver has an output voltage versus current characteristic which saturates at

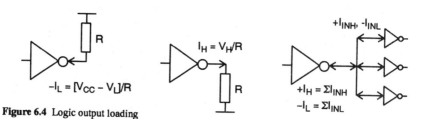

Figure 6.4 Logic output loading

some level of loading (Figure 6.5). The characteristic is tailored so that at a given load current, the output voltage V_{OH} or V_{OL} is equivalent to the input threshold voltage (V_{IH} or V_{IL}) plus the noise immunity for that particular logic family. This load current then

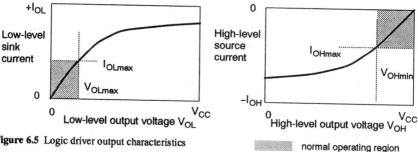

Figure 6.5 Logic driver output characteristics   ▨ normal operating region

corresponds to the sum of the input currents for a given number N of standard gates of that family, and N is called the "fan-out": that number of standard gates that the output can drive and still keep the interface within the noise threshold limits. The fan-out is normally specified against each output of a device for other devices of the same family, but it can be calculated for other logic family interfaces simply by comparing the output voltage and current capability for each logic state with the current and voltage requirements of each input. As before, fan-out figures for logic high and low may differ.

Example: MC68HC11 port driving several 74LS devices

Assume that, to expand the use of the 68HC11's I/O ports, you need to put three octal latches (74LS373) on one 8-bit port together with a 3-to-8 decoder (74LS138) and a NAND gate (74LS00).

The loading on the port pins will be three, four or five LS-TTL inputs. Each input is rated at –400µA I_{IL} and 20µA I_{IH}. Thus five inputs will take –2mA low and 100µA high, while four will take –1.6mA low and 80µA high.

The 68HC11 outputs are specified as 3.7V V_{OH} at –0.8mA load and 0.4V V_{OL} at 1.6mA load. The V_{OH} specification presents no problem, but attaching a logic 0 load of –2mA will take V_{OL} higher than 0.4V by an unspecified amount, and will degrade the noise immunity.

Four inputs are just within spec., so the fan-out for this interface is 4.

You can get over this by changing to 74HC throughout, or if you must use 74LS373's, by changing either or both of the other two devices to 74HC. Alternatively, re-design your port circuitry so that the NAND gate does not share outputs with the decoder.

Normally, at least 10 devices of the same family can be driven from one output and this is enough for most purposes. When low-power parts are required to drive high-power types then fanout may be insufficient and you have to use intermediate buffer devices. The low drive capability of many microprocessor bus outputs severely restricts the number of other components that may be placed on the bus without interposing additional bus buffers – which accounts for the phenomenal popularity of the 74XX244 type of octal bus driver!

Dynamic loading

The dc load current taken by the input side of the interface is only part of the total load. Indeed, for CMOS-input logic ICs it is negligible and has no significant effect on fanout calculations. But every input, CMOS, TTL or any other, has an associated capacitance to ground and the charging or discharging of this capacitance limits the speed at which the node can operate. Typical logic IC input capacitances are 5–10pF and these must be summed for all connected inputs, together with an allowance for interconnection capacitance which is layout-dependent but is again typically 5–10pF, to reach the total load capacitance facing the driver.

Driver dynamic output current ability is rarely specified on data sheets but some manufacturers give application guidance. For instance, the 74HC range can offer typically ±40mA for standard devices and ±60mA for buffers at 4.5V supply. This current slews the interface node capacitance C_n (Figure 6.6) that you have just calculated and you have to ensure that slewing from logic 0 to the logic 1 threshold, or vice versa, is accomplished before the input data level is required to be valid. As an

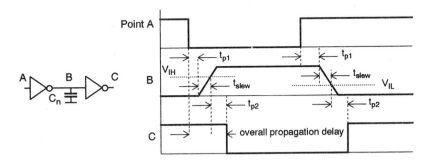

Figure 6.6 Propagation and slewing delays

example, a 100pF capacitance slewing from 0 to 3V with 40mA drive will take 7.5ns, and this time (plus a safety factor) has to be added onto the other specified propagation delays to ensure adequate timing margin. If the figures don't add up, you will need to add extra buffer devices (which add their own propagation delays), reduce the load, reduce the operating frequency or go to a faster logic family.

If you choose to run NMOS or CMOS devices with a high load capacitance and accept that the edges will be slower, then be aware that this also reduces the reliability of the device because of the higher transient currents that the output drivers are handling.

6.1.3 Induced switching noise

This phenomenon is more colloquially known as "ground bounce". We are not talking here about external noise signals, but about noise which is induced on the supply rails by the switching action of each logic gate in the circuit.

As each gate changes state, a current pulse is taken from the supply pins because of the different device currents required in each state, the external loading, the transient caused by charging or discharging the node capacitance, and the conduction overlap in the totem-pole output stage. All these effects are present in all logic families to some extent, although CMOS types suffer little from the first two. In most cases, the node capacitance charging current dominates, more so with higher-speed circuits. The capacitance C_n must be charged with a current of

$$I \quad = \quad C_n \cdot dV/dt$$

Thus a 74ALS-series gate with a dV/dt of 1V/ns will require a 50mA current pulse when charging a 50pF node capacitance. Figure 6.7 shows the current paths. The significance of the supply current spike is that it causes a disturbance in the supply voltage and also in the ground line, because of the inductive reactance of the lines. A pulse with a di/dt of 50mA/ns through a track inductance of 20nH (one inch of track) will generate a voltage pulse of 1V peak, which is approaching the noise margin of fast logic. Supply voltage spikes are not too much of a problem as the logic high level noise immunity is usually good and they can be attenuated by proper decoupling, as the next section shows. Ground line disturbances are more threatening. Pulses on a high-impedance ground line can easily exceed the noise threshold and cause spurious switching of innocent gates. Only if a good, low-inductance ground system is maintained, as discussed in chapter 1 and again in section 2.2.4, can this problem be minimised.

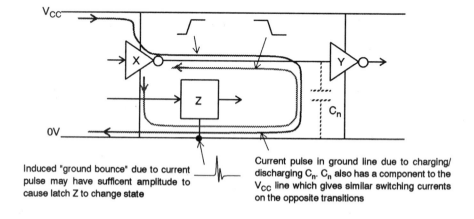

Induced "ground bounce" due to current pulse may have sufficent amplitude to cause latch Z to change state

Current pulse in ground line due to charging/ discharging C_n. C_n also has a component to the V_{CC} line which gives similar switching currents on the opposite transitions

Figure 6.7 Induced ground noise due to switching currents

Synchronous switching

The supply pin pulse current is magnified in synchronous systems when several gates switch simultaneously. A typical example is an octal bus buffer or latch whose data changes from #FF$_H$ to #00$_H$. If all outputs are heavily loaded, as may be the case when the device is driving a large data bus, a formidable current pulse – exceeding an amp in fast systems – will pass through the ground pin. Worse, if seven bits of an octal latch change simultaneously, the induced ground bounce may corrupt the state of the eighth bit. You need to ensure that such devices are grounded to their loads with a very low-inductance ground system, preferably a true ground plane.

Ground noise on a microprocessor board can easily be observed by hooking a wide-bandwidth oscilloscope to the ground line – you can connect the 'scope probe tip and its ground together and still see the noise, since the probe lead braid is a high enough impedance at these frequencies to convert what should be a common-mode signal to differential mode. What you see is a regular series of narrow, ringing pulses spaced at the clock period. The amplitude of each pulse varies because the sum of the data transitions is random, but the timing does not. Such noise (Figure 6.8) is impossible to remove entirely.

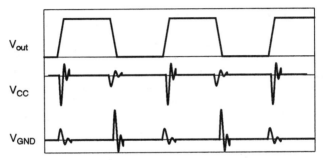

Figure 6.8 Switching noise on power and ground rails

6.1.4   Decoupling

No matter how good the V$_{CC}$ and ground connections are, you cannot eliminate all line inductance. Except on the smallest boards, track distance will introduce an impedance which will create switching noise from the transient currents discussed in the last section. This is the reason for decoupling (Figure 6.9).

Distance

The purpose of a decoupling capacitor is to maintain a low dynamic impedance from the individual IC supply voltage to ground. This minimises the local supply voltage

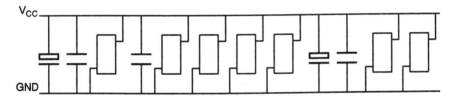

Figure 6.9 Logic decoupling scheme

droop when a fast current pulse is taken from it. The word "decoupling" means isolating the local circuit from the supply impedance. Bearing in mind the speed of the current pulses just discussed, it should be clear that the capacitor must be located close to the circuit it is decoupling. "Close" in this context means less than half an inch for fast logic such as AS/ALS-TTL or ECL, especially when high current devices such as bus drivers are involved, extending to several inches for low-current, slow devices such as 4000B-series CMOS.

If the decoupling current path between IC and capacitor is too long, the track inductance in conjunction with the capacitor forms a high-Q LC tuned circuit, and the ringing it generates will have a worse effect than no decoupling at all.

Capacitor type and value

The crucial factor for high-speed logic decoupling is lead inductance rather than absolute value. Figure 3.17 is instructive in this regard. Minimum lead inductance offers a low impedance to fast pulses. Small disk or multilayer ceramics, or polyester film types, are preferred; chip capacitors are even better.

You can calculate the value if you want to by matching the transient current demand to the acceptable power rail voltage droop. Take for example a 74HC octal buffer each of whose outputs when switching takes a transient current of 50mA for 6ns (calculated from $I = C_n \cdot dV/dt$). The total peak current demand is then 0.4A.

The acceptable voltage droop is perhaps 0.4V (equivalent to the worst system noise margin). Assume that the local decoupling capacitor supplies all of the current to hold the droop to this level, which is reasonable since other decoupling capacitors on the board will be isolated by track inductance. Then the minimum capacitor value is

$$C \quad = \quad I \cdot t/V \quad = \quad 0.4 \cdot 6 \cdot 10^{-9} / 0.4 \quad = \quad \underline{\mathbf{6nF}}$$

On the other hand, the actual value is non-critical, especially as the variables in the above calculation tend to be somewhat vague, and you will prefer to use the same component in all decoupling positions for ease of production. Values between 10 and 100nF are recommended, a good compromise being 22nF, which has both low self-inductance and respectable reservoir capacitance. It also tends to be cheaper than the higher values, particularly in the low-performance Z5U ceramic grade, which is adequate for this purpose.

Capacitors under the IC package

Very high-speed and high-current logic ICs push the location requirement of the decoupling capacitor to its limits: it has to be right next to the supply pins. In fact, the inductance of the lead-out wires within the chip package becomes significant, and this has led to a proposal to locate power and ground pins in the middle of the package rather than at opposite corners for certain high-performance ICs. Be that as it may, the ultimate in decoupling is to locate the capacitor underneath the chip, between it and the board.

Such flat capacitors are available, matched to the various dual-in-line package sizes and sharing holes with the IC power and ground pins. They can be obtained either as discrete components or integrated with the IC socket, and can offer an intrinsic self-inductance of 2nH or less, with an improvement in decoupling effectiveness which is most marked at frequencies above 50MHz. Their main disadvantage is that they are relatively expensive and not widely sourced as yet.

Low-frequency decoupling

You also need to decouple the supply rail against lower-frequency ripple due to varying logic load currents, as distinct from transient switching edges. The frequency of these ripple components is in the megahertz range and lower, so that widely distributed capacitance and low self-inductance are less important. Typically, they can be dealt with by a few tantalum electrolytics of $1 - 2\mu F$ placed around the board, particularly where there are several devices that can turn on together and produce a significant drain from the power supply, such as burst refresh in dynamic RAM. Additionally, a single large capacitor of $10 - 47\mu F$ at the power entry to the board is recommended to cope with frequency components in the kHz range.

Under normal circumstances, logic circuits are inherently insensitive to ripple on the supply lines. The exception is when they are faced with slow edges; if the ripple is at a substantially higher frequency than the edge and is modulating the signal, then as the signal passes through the transition region the logic element may undergo spurious switching (Figure 6.10).

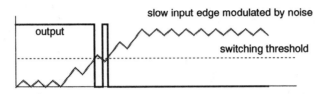

Figure 6.10 Spurious switching on slow edges due to ripple

The safest way to deal with slow edges is with a Schmitt-trigger logic input.

Guidelines

The minimum requirements for good decoupling are:

- one $22\mu F$ bulk capacitor per board
- one $1\mu F$ tantalum capacitor per 10 packages of SSI/MSI logic or memory
- one $1\mu F$ tantalum capacitor per 2–3 LSI packages
- one 22nF ceramic or polyester capacitor for each octal IC or for each MSI/LSI package
- one 22nF ceramic or polyester capacitor per 4 packages of SSI logic

When in doubt, calculate the requirements for individual power/speed-hungry devices to make sure you have enough capacitors, and that they are in the right places.

6.1.5   Unused gate inputs

Frequently you will have spare gates or latches in a package left over, or will not be using all the inputs of a multi-input gate or latch. All such unused logic inputs must be tied to a fixed voltage, either high or low, and should never be left floating. There is an exception to this: TTL inputs, if left open, will assume a logic high state, and inputs of unused gates can be allowed to float if no use is made of their outputs. Noise immunity of floating inputs is poor, so you should not float spare inputs of used gates, and especially not preset/clear inputs of latches or flip-flops, which are very sensitive to spikes.

The rules for connecting unused inputs are different for TTL and CMOS logic.

Figure 6.11 illustrates the options.

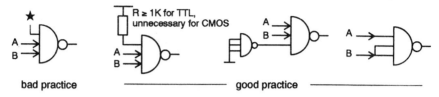

Figure 6.11 Connecting unused inputs

TTL families

Do not tie inputs directly to the V_{CC} line. TTL inputs are susceptible to damage from transient overvoltages of more than 5.5V, which can occur surprisingly often even on well-decoupled supplies. Instead, tie them to V_{CC} through a resistor of at least 1kΩ or, better still, drive them from the output of an unused gate held at logic 1. If there is spare fanout capacity, common unused inputs to another input on the same gate and drive them. This results in optimum speed since all stray capacitance associated with the gate inputs is being driven from a low impedance. Two inputs paralleled on the same gate have a fan-in of 1 in the zero state, and may be taken as 1.5 in the high state.

Connecting an input to ground is permissible, but takes extra dc current drain due to I_{IL}. Most MSI devices need their inputs to be taken high if they are not active.

CMOS families

You must connect *all* unused CMOS inputs either to V_{CC} or ground. Floating any input is inadmissible, whether its gate is used or not. This is because the CMOS input has a very high impedance and consequently can float to any voltage if unconnected, and this voltage could be within the threshold switchover region of the gate. At this point both the P-channel and N-channel input transistors are conducting, which results in excessive current drain through the package. Due to the high gain of buffered gates, it is possible for a gate to oscillate, resulting in even higher current drain.

CMOS inputs can be connected directly to either rail; a protection resistor is unnecessary.

6.2 Interfacing

6.2.1 Mixing analogue and digital

The two main problems which face designers who have to integrate analogue and digital circuits on the same pcb are

- preventing digital switching noise from contaminating the analogue signal, and
- interfacing the wide range of analogue input voltages to the digital circuit.

Generating analogue outputs from digital signals is not usually a problem. Generating digital inputs from analogue signals is.

Ground noise

The high-frequency switching noise that was discussed in section 6.1.3 must be kept out of analogue circuits at all costs. An analogue-to-digital interface quantizes a variable analogue signal into a digital word, and the number of bits in the word determines the resolution that can be achieved of the signal. Assuming a full-scale voltage range of 0 to 10V, which is typical of many analogue-digital converters (ADCs), Table 6.2 shows the voltage levels that correspond to one bit change in the digital word.

Word length	Resolution voltage
8 bit	39mV
10 bit	10mV
12 bit	2.4mV
14 bit	0.6mV
16 bit	0.15mV

Table 6.2 ADC resolution voltage for different word lengths, 10V full-scale

You can see that the more resolution is demanded of the interface, the smaller the voltage change that will cause one bit change. 8 bits is regarded as commonplace in ADC circuits, 12 bits as reasonably high resolution (0.025%) and 16 bits as precision.

The significance of these diminishing voltage levels is that any noise that is coupled into the analogue input will cause unwanted fluctuation of the digital value. For a 12-bit converter, a one-bit uncertainty will be given by noise of 2.4mV at the converter input; for a 16-bit, this reduces to 150 microvolts. By contrast, the switching noise on the digital ground line is normally tens of millivolts and frequently hundreds of millivolts peak amplitude. If this noise were coupled into the converter input – and it is hard to keep ground noise out of the input – you would be unable to use a converter of greater precision than 8-10 bits.

Filtering

One partial solution to this problem is to filter the bandwidth of the analogue signal to well below that of the noise so that the effective noise signal is reduced. For slowly-varying analogue signals this works reasonably well, especially if the noise injection occurs at the input of the signal-processing amplifier so that bandwidth limitation has maximum effect. Filtering is in any case good practice to minimise susceptibility to external noise.

Filtering the input amplifier is no use if the noise is injected into the ADC itself. For fast ADCs and wide-bandwidth analogue signals you cannot take this approach anyway and the only available solution is to prevent the injection of digital noise at its source.

Segregation

The basic rule to follow when designing an analogue-to-digital interface is to segregate the circuits, including grounds, completely. This means that

- separate analogue and digital grounds should be established, connected only at one point
- the analogue and digital sections of the circuit should be physically separated, with no digital tracks traversing the analogue section or vice versa. This will minimise crosstalk between the circuits.

Single-board systems

The appropriate grounding schemes for single-board and multi-board systems are shown in Figure 6.12. If your system has a single analogue-to-digital converter, perhaps

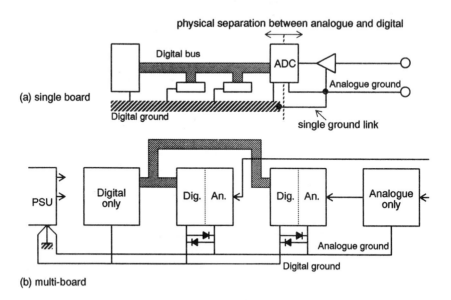

Figure 6.12 Ideal layout for analogue and digital grounds

with a multiplexer to select from several analogue inputs, then the connection between analogue and digital grounds can be made at this ADC as in Figure 6.12(a). This scheme requires that the analogue and digital power supply returns are not linked together anywhere else, so that two separate power supply circuits are needed. The analogue and digital grounds must be treated as entirely separate tracks, despite being nominally at the same potential; unavoidable noise currents circulating in the digital ground will then not couple into the "clean" analogue ground. The digital ground should be of gridded or ground plane construction, whereas the analogue section will normally benefit from a single-point grounding system. On no account should you extend the digital ground plane over the analogue section of the board.

Multi-board systems

When your system consists of several boards, some entirely digital, some entirely analogue and some a mixture of the two, with an external power supply, then you cannot make the connection between digital and analogue grounds at the ADC. There may be several ADCs in the one system. Instead, make the link at the power supply (Figure 6.12(b)) and run separate analogue and digital grounds to each board that requires them. Digital-only boards should be located physically closer to the power supply to minimise the radiating loop area or length. Where analogue and digital grounds are both taken to one board then follow the segregation rules outlined above, and include clamping diodes between the two grounds to guard against accidental disconnection of one or the other.

6.2.2   Generating digital levels from analogue inputs

The first rule when you want to use a varying analogue voltage to generate an on/off digital signal – as distinct from an analogue-to-digital conversion – is: always use either a comparator or a Schmitt-trigger gate. *Never* feed an analogue signal straight into an ordinary TTL or CMOS gate input.

The reason is that ordinary gates do not have well-defined input voltage switching thresholds. Not only that, but they are also very critical of slow rise-time inputs. Very few analogue input signals have the slew rate, typically faster than 5V/μs, required to produce a clean output from an ordinary logic gate. The result of applying a slow analogue voltage to a logic gate is shown in Figure 6.12.

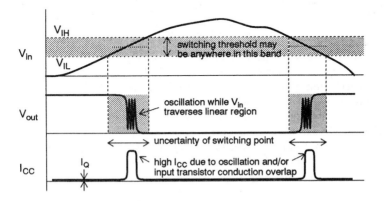

Figure 6.12 The effect of a slow input to a logic gate

A Schmitt trigger gate, or a comparator with hysteresis (see section 5.3.4), will get over the slow rise-time problem. A Schmitt trigger gate has the same output characteristics as an ordinary gate but it includes input hysteresis to ensure a fast transition. The threshold levels of typical Schmitt devices, such as the 74HC14, are specified within wide tolerances and so do not overcome the variability of the actual switching point. When the analogue levels corresponding to high and low states can be kept above V_{IH} and below V_{IL} respectively, a Schmitt is adequate. For more precision you will need to use a comparator with an accurately specified reference voltage.

Secondly, if the analogue supply rail range is greater than the logic supply, interfacing the analogue signal straight to the logic input will threaten the gate with damage. This is possible even if the normal signal range is within the logic supply range; abnormal conditions such as turn-on or turn-off may exceed the rails. This, of course, is also a problem with Schmitt trigger gates. Normally, the inputs are protected by clamp diodes to the supply and ground rails, but the current through these must be limited to a safe level so a resistor in series with the input is essential. More positive steps to limit the input voltage, such as running the analogue section from the same supply voltage as the logic (heeding the earlier advice about separate digital and analogue ground rails), are to be preferred.

De-bouncing switch inputs

On the face of it, switch inputs to digital circuitry must be the easiest of interfaces. All you should need are an input port or gate, a pull-up resistor and a single pole switch

(Figure 6.14). This circuit, though it undoubtedly works, is prone to a serious problem because of the electromechanical nature of the switch and the speed of logic devices.

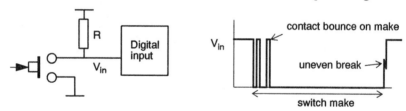

Figure 6.14 Contact bounce

When a switch contact operates, the current flow is not cleanly initiated or interrupted. As the contacts come together or part, the instantaneous contact resistance varies due to contamination, and the mating surfaces may "bounce" apart a few times due to the springiness of the material. As a result the switching edge is irregular and may easily consist of several discrete edges, extending over a period of typically 1ms. You can verify this behaviour simply by observing the input waveform of Figure 6.14 on a storage 'scope.

Of course, the digital input responds very fast to each crossing of the switching threshold, and consequently the port or gate sees several transitions each time the switch is operated, before it settles to a steady-state 1 or 0. This may not be a problem for level-sensitive inputs, but it undoubtedly is for edge-sensitive ones such as counter or latch clock inputs. Mis-triggering of counter circuits that are fed from a switch input is commonly caused by this phenomenon.

The simple solution to contact bounce is to filter the logic input with an RC network (Figure 6.15(a)). The RC time constant must be significantly longer than the bounce period to effectively attenuate the contact noise. This has the extra advantage of protecting against induced impulsive or RF interference, but it requires additional discrete components and demands that the logic input must be a Schmitt-trigger type, since the input rise-time has been deliberately slowed. It is also not well suited to TTL-family inputs, which take a comparatively high input current and cannot tolerate much R_{in}.

If the switch input may change state quickly, an RC time constant which is sufficiently long to cure the bounce will slow the response to the switch unacceptably. This can be overcome in two ways: the R-S latch, Figure 6.15(b), which requires a changeover rather than single-throw switch, or a software- or hardware-implemented delay. Figure 6.15(c) shows the hardware delay, which uses a continuously-clocked shift register and OR gate to effectively "window out" the bounce. The delay can be adjusted to suit the bounce period. These two solutions are most suited to realisation with semi-custom logic arrays or ASICs, where the overhead of the extra logic is low.

6.2.3   Protection against externally-applied overvoltages

Logic inputs and outputs which are taken off-board will be subject at some time in the life of the system to an overvoltage. Your philosophy in this respect should be, if it can happen, it will. Overvoltages can be applied by misconnection of the board or of external equipment, or can be due to static build-up. The latter is a particular threat to CMOS inputs with their high impedance, but the effect of a large static discharge can also be disastrous for other logic families.

There are three major consequences of an overvoltage on a logic signal line:

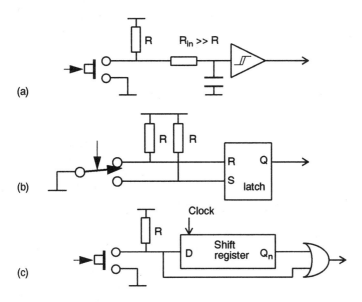

Figure 6.15 Switch de-bouncing circuits

- immediate damage of the device due to rupture of the track metallization or destruction of the silicon;

- progressive degradation of device characteristics when the overvoltage does not have enough energy to destroy it immediately;

- latch-up, where damage may be caused by excessive power supply current following a transient overvoltage.

Modern logic families include some protection both at their inputs and outputs in the form of clamping diodes to the supply lines, but these diodes are limited in their current-handling ability and therefore the potential fault current that can be applied due to an overvoltage must be limited. This is best achieved by the methods of Figure 6.16.

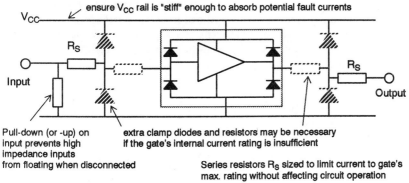

Figure 6.16 Logic gate I/O protection

6.2.4   Isolation

Even if you take precautions against input/output abuse, it is not good practice to take logic signals directly into or out of equipment. As well as facing the threat of overload on individual lines, you also have to extend the ground and/or supply rails outside the equipment to provide a signal return path. These then act as antennas both to radiate ground noise out of the equipment and to conduct external interference back into it. It is very much safer to keep power rails within the bounds of the equipment case.

A common technique to achieve this is to electrically isolate all signal lines entering or leaving the equipment. As well as guarding against interference, this eliminates problems from ground loops and ground differentials. Digital signals lend themselves to the use of opto-couplers. An opto-coupler is basically an LED chip integrated in the same package as a light-sensitive device such as a photodiode or phototransistor, the two components being electrically separate but optically coupled. A typical isolation scheme using such devices is shown in Figure 6.17.

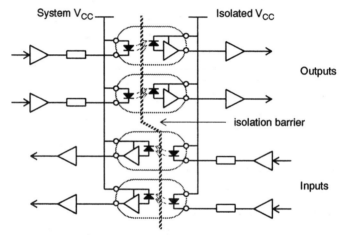

Figure 6.17 Interface isolation using opto-couplers

One opto-coupler is needed per digital channel. Opto-couplers can be sourced as single, dual or quad packages, and the price for commercial grade units can vary depending mainly on the required speed and level of integration from 25p to £5 per channel. Clearly, in cost- or space-sensitive applications the number of isolated channels should be minimised. This tends to mean that isolation is more common in industrial products than consumer ones.

Opto-coupler tradeoffs

There are a number of quite complex tradeoffs to make when you use opto-isolation. Factors to consider are

- speed of the interface. Cheap couplers with standard transistor outputs have switching times of 2–5µs so are limited to data rates of around 100kbits/s maximum. High speed devices with data rates of 10Mbits/s are available but cost over £5 per channel.

- power consumption. Standard transistor output types offer a current transfer ratio (CTR) typically between 10 and 80%. This is the ratio of LED input current to transistor output current in the on-state. Thus for a required output current of 1mA with a CTR of 20%, 5mA would be needed through the LED. Also, CTR degrades with time and you should include an extra safety margin of between 20 and 50%, depending on expected lifetime and operating current, to ensure end-of-life circuit reliability. Reducing the operating current reduces the speed of the interface. Darlington-output opto-couplers are available with CTRs of 200–500%, but these unfortunately have turn-off times of around 100µs so are only useful for low-speed purposes. Opto-coupler drive current can be a significant fraction of the overall power requirement, especially on the isolated side.

- support circuitry. A simple photo-transistor or photó-darlington output needs several passive components plus a buffer gate to interface it correctly to logic levels. Alternatively, you can get opto-couplers which have logic-compatible inputs and outputs, especially the faster ones, but at a significantly higher cost. Low-current LED drive requirements can be met directly by a logic gate and series limiting resistor, whereas if you are using a cheaper opto with higher LED current you will need an extra buffer.

Coupling capacitance

Although an opto-coupler breaks the electrical connection at dc, with an isolation voltage measured in kV, there is still some residual coupling capacitance which reduces the isolation at high frequencies. The specification figures of 0.5–2pF are increased somewhat by stray wiring capacitance which is layout dependent. Input and output pins are invariably on opposite sides of the package. There is no point in designing in an opto-coupler for isolation if you then run the output tracks back alongside the input tracks!

The coupling capacitance of individual channels, multiplied by the number of channels in the system, means that a significant level of high frequency ground noise may still be coupled out of an isolated system, or fast-risetime transients or RFI may still be coupled in. (This is another argument for minimising the number of channels.) Also, high common-mode dV/dt signals can be coupled directly into the photo-diode or transistor input through this capacitance and cause false switching. This effect is reduced by incorporating an electrostatic screen across the optical path and connecting it to the output ground pin, and some optocouplers are available with this screen included. Common-mode transient immunity can vary from worse than 100V/µs to better than 5kV/µs (for the expensive devices).

Alternatives to opto-couplers

Two alternatives to opto-couplers for isolating digital signals are relays and pulse transformers. The relay is a well-established device and is a good choice if its restrictions of size, weight, speed, power consumption and electromechanical nature are acceptable.

Pulse transformers are most useful for passing wide bandwidth, high speed digital data for which opto-couplers are too slow or too expensive. They can also be designed for good immunity to high dV/dt interference. The data must be coded or modulated to remove any dc component. This requires an overhead of a few gates and a latch per channel, but this overhead may be acceptable, especially if you are already using semi-custom silicon, and may easily be outweighed by the attractions of high speed and low power consumption.

6.2.5   Data interface standards

When you want to connect logic signals from one piece of equipment to another, it is not sufficient to use standard logic devices and make direct gate-to-gate connections, even if they are isolated from the main system. Standard logic is not suited to driving long lines; line terminations are unspecified and noise immunity is low, so that reflections and interference would give unacceptably high data corruption. External logic interfaces must be specially designed for the purpose.

At the same time, it is essential that there is some commonality of interface between different manufacturers' equipment. This allows the user to connect, say, a computer from manufacturer A to a printer from manufacturer B without worrying about electrical compatibility. There is therefore a need for a standard definition for electrical interface signals.

This need has been recognised for many years, and there are a wide variety of data interchange standards available. The logic of the marketplace has dictated that only a small number of these are dominant. This section will consider the two main commercial ones: EIA-232D and EIA-422. These were published by the US Electronic Industries Association in 1987 and 1977 respectively. EIA-232D is an update of the popular RS-232C standard published in 1969, to bring it into line with the international CCITT V.24 and V.28 and ISO IS2110 standards. EIA-422 is the same as the earlier RS-422 standard. The prefix changes are cosmetic, purely to identify the source of the standards as the EIA.

EIA-232D

The boom in data communications has led to many products which make interface conformity claims by quoting "RS232" in their specifications. Some of these claims are in fact quite spurious, and discerning users will regard interface conformity as an indicator of product quality, and test it early on in their evaluation. The major characteristics of the specification are given in Table 6.3. As well as specifying the electrical parameters, EIA-232D also defines the mechanical connections and pin configuration, and the functional description of each data circuit.

By modern standards the performance of EIA-232D is primitive. It was originally designed to link data terminal equipment (DTE) to modems, known as data communications equipment (DCE). It was also used for data terminal-to-mainframe interfaces. These early applications were relatively low speed, less than 20kbaud, and used cables shorter than 50 feet. Applications which call for such limited capability are now abundant, hence the standard's great popularity. Its new revision recognises this by replacing the phrase "data communication equipment" with "data circuit-terminating equipment", also abbreviated to DCE. It does not clarify exactly what is a DTE and what is a DCE, and since many applications are simple DTE (computer) to DTE (terminal or printer) connections, it is often open to debate as to what is at which end of the interface. The common sight of an installation technician crouched over a 25-way connector swapping pins 2 and 3, to make one end's receiver listen to the other end's driver, will not disappear.

EIA-232D's transmission distance is limited by its unbalanced design and restricted drive current. The unbalanced design is very susceptible to external noise pick-up and to ground shifts between the driver and receiver. The limited drive current means that the slew rate must be kept slow enough to prevent the cable becoming a transmission line, and this puts a limit on the fastest data rate that can be accommodated. Maximum cable length, originally fixed at 50 feet, is now restricted by a requirement for maximum load capacitance (including receiver input) for each circuit of 2500pF.

Note that there are several common "enhancements" that are not permitted by strict adherence to the standard. EIA-232D makes no provision for tri-stating the driver output, so multiple driver access to one line is not possible. Similarly, paralleling receivers is not allowed unless the combined input impedance is held between 3kΩ and 7kΩ. It does not consider electrically isolated interfaces: no specification is offered for isolation requirements, despite their desirability. It does not specify the communication data format. The usual "one start bit, eight data bits, two stop bits" format is not part of the standard, just its most common application. It is not directly compatible with another common single-ended standard, RS-423, although such connections will usually work. Also, you cannot run EIA-232D off a ±5V supply rail – the minimum driver output voltage is specified as ±5V, loaded with 3-7K and with an output impedance of 300Ω.

EIA-422

Many data communications applications now require data rates in the megabaud region, for which EIA-232D is inadequate. This need is fulfilled by the EIA-422 standard, which is an electrical specification for drivers and receivers for use in a balanced or differential, point-to-point high speed interface using twisted-pair cable. Table 6.3 summarizes the EIA-422 specification in comparison with EIA-232D. One driver and up to ten receivers are allowed. The maximum data rate is specified as 10Mbaud, with a tradeoff against cable length; maximum cable length at 100kbaud is 4000 feet. Note that unlike EIA-232D, EIA-422 does not specify functional or mechanical parameters of the interface. These are included in other standards which

Interface	EIA-232D	EIA-422
Line type	Unbalanced	Balanced, differential
Line impedance	Not applicable	100Ω
Max. line length	Load dependent, typically 50ft	4000ft at 100kbaud reducing to 40ft at 10Mbaud
Max. data rate	20kbaud	10Mbaud
Driver		
Output voltage	±5 to ±15V loaded with 3–7kΩ	±6V max unloaded, ±2V min loaded with 100Ω
Output impedance	Not specified	50Ω max
Short circuit current	500mA max	150mA max
Max. rise time	4% of unit interval 1ms–5μs 30V/μs max. slew rate	10% of unit interval
Output with power off	> 300Ω output resistance	±100μA max. leakage current
Receiver		
Sensitivity	±3V max. thresholds	200mV max
Input impedance	3kΩ – 7kΩ, < 2500pF	4kΩ min
Common-mode voltage	Not applicable	±7V

Table 6.3 Major electrical characteristics of EIA-232D and EIA-422

incorporate it, notably EIA-449 and EIA-530.

EIA-422 achieves its high-speed and long-distance capabilities by specifying a balanced and terminated design. The balanced design reduces sensitivity to external common-mode noise and allows a ground differential of up to a few volts to exist between the driver and one or more of the receivers without affecting the receiver's thresholds. A cable termination, together with increased driver current, allows fast slew rates which in turn allows high data rates. If the cable is not terminated, serious ringing on the edges occurs which may cause spurious switching in the receiver. The specified termination of 100Ω is closely matched to the characteristic impedance of typical twisted-pair cables.

Interface design

By far the easiest way to realise either EIA-232D or EIA-422 interfaces is to use one of the many specially-tailored driver and receiver chip sets that are available. The more common ones, such as the 1488 driver/1489 receiver for EIA-232D or the 26LS31 driver/26LS32 receiver for EIA-422, are available competitively from many sources and in low-power CMOS versions. You can also obtain combined driver/receiver parts so that a small interface can be handled with one IC. The high-voltage requirement of EIA-232D, typically ±12V supplies, is addressed by some suppliers who offer on-chip dc-to-dc converters from the +5V rail.

Figure 6.18 suggests typical interface circuits for the two standards. Note the inclusion of power supply isolating diodes, to protect the rest of the circuit against the inevitable overvoltages that will come its way. You can also construct an interface, particularly the simpler EIA-232D, using standard components such as op-amps, comparators, CMOS buffer devices or discrete components if you are prepared to spend

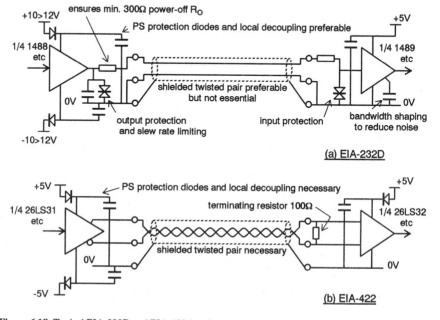

Figure 6.18 Typical EIA-232D and EIA-422 interface circuits

some time characterising the circuit against the requirements of the standard and against expected overload conditions. This may turn out to be marginally cheaper in component cost, but its overall worth is somewhat questionable.

6.3   Microprocessor watchdogs and supervision

Microprocessors are exceptionally versatile devices. They can be applied to virtually any control, processing or data acquisition task. But they are not entirely self-contained: like children, they need care and attention, and occasional corrective action, if they are to function reliably and properly.

6.3.1   The threat of corruption

A microprocessor is a state machine. It steps predictably from state to state, its operation controlled entirely by the contents of the program counter and program memory. Provided that these are never misinterpreted, it will follow its program correctly and, assuming its software has been fully tested, will never deviate from its operational specification. If the digital circuit design rules outlined previously have been followed, there is no intrinsic reason why the data should be misinterpreted.

In the real world, though, there *is* a mechanism by which data gets corrupted. This is when an external electromagnetic influence is coupled onto the signal, clock or supply lines and over-rides the transfer of stored program data (Figure 6.19). The most usual threat is from a brief, sub-microsecond transient due to a nearby electrostatic discharge or coupled in via mains or signal cables, but strong continuous or pulsed

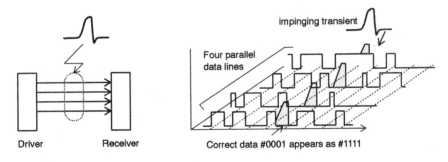

Driver   Receiver   Correct data #0001 appears as #1111

Figure 6.19 Transient affecting data transfer

radio-frequency fields can have the same effect. Nearly all microprocessor circuits operate with clock frequencies and data rates in excess of 1MHz, some much higher. Possibly only one data bit need be corrupted to derail the program completely, and this can be achieved with a transient of only a fraction of a microsecond duration.

Chapter 8 deals with the circuit techniques to minimise the amplitude of such disruptive interference, and these go a long way towards "hardening" a microprocessor circuit against corruption. But they cannot *eliminate* the risk. The coincidence of a sufficiently high-amplitude transient with a vulnerable point in the data transfer, both in time and in the three dimensions of the pcb layout, is an entirely statistical affair. The most cost-effective way to ensure the reliability of a microprocessor-based product is to accept that the program *will* occasionally be corrupted, and to provide a means whereby the program flow can be automatically recovered, preferably transparently to the user. This is the function of the microprocessor watchdog.

Power rail supervision

The other significant times when microprocessor operation may go astray is during the transitions of its power supply. All micros are characterised for a stable power rail of typically 4.75 to 5.25V, though some allow wider tolerances. They have no guaranteed behaviour below the specified minimum voltage, and so their operation while the power rail is ramping up or collapsing is an unknown quantity. Conventionally, the case of power-on has been treated by delaying the release of the micro's RESET input. It may seem that if the power has been switched off it doesn't much matter what the micro does, but the problem has been enormously compounded by the introduction of non-volatile memory and real-time clocks. These should offer guaranteed security of data over power cycling if they are to be worth including, and so the micro must be effectively prevented from corrupting them under abnormal power conditions.

These considerations have led to the development of techniques for power rail supervision, including the functions of power-up and power-down reset, power fail detection, and write protection and battery back-up switching for non-volatile memory. They will be briefly touched on in the context of power supplies in chapter 7; here we shall discuss their application at the processor end.

6.3.2   Watchdog design

Some of the more up-to-date micros on the market include built-in watchdog devices, which may take the form of an illegal-opcode trap, or a timer which is repetitively reset by accessing a specific register address. These are more common in single-chip microcontrollers such as the Motorola 68HC11. This is an excellent example of IC designers listening to their customers, since early micros had no watchdog provision and circuit designers had to learn the hard way about the need for one.

If your chosen micro has an on-board watchdog, use it without hesitation. It will be closely matched to the processor's peculiarities and requirements. If it hasn't, read on.

Basic operation

The most serious result of a transient corruption is that the processor program counter or address register is upset, so that it starts interpreting data or empty memory as valid instructions. This causes the processor to enter an endless loop, either doing nothing or performing a few meaningless instructions. A similar effect can happen if the stack register or memory is corrupted. Either way, the processor will appear to be catatonic, in a state of "dynamic halt".

A watchdog guards against this eventuality by requiring the processor to execute a specific simple operation regularly, regardless of what else it is doing, on pain of consequent reset. The watchdog is actually a timer whose output is linked to the RESET input, and which itself is being constantly retriggered by the operation the processor performs, normally writing to a spare output port. This operation is shown schematically in Figure 6.20.

Timeout period

If the timer does not receive a "kick" from the output port for more than its timeout period, its output goes low ("barks") and forces the microprocessor into reset. Clearly the timeout period is an important system parameter. It must be long enough so that servicing the port does not require the processor to interrupt time-critical tasks, and so

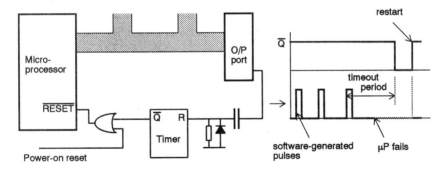

Figure 6.20 Watchdog operation

that there is time for the processor to start the servicing routine when it comes out of reset (otherwise it would be continually barking and the system would never restart properly). On the other hand, it must not be so long that the operation of the equipment could be corrupted for a dangerous period. There is no one timeout period which is right for all applications, but usually it is somewhere between 10ms and 1s.

Timer hardware

The watchdog circuit has to exceed the reliability of the rest of the circuit and so the simpler it is, the better. A standard timer IC such as the 555 or its derivatives is quite adequate. However, the 555 timeout period is set by an RC time constant and this may give an unacceptably wide variation in tolerance, besides needing extra discrete components. A digital divider such as the 4060B fed from a high-frequency clock and periodically reset by the report pulses may be a more attractive option, since no other components are needed. The clock has to have an assured reliability in the presence of transient interference, but such a clock may well already be present or could be derived from the unsmoothed AC input at 50/60Hz.

An extra advantage of the digital divider approach is that its output in the absence of retriggering is a stream of pulses rather than a one-shot. Thus if the micro fails to be

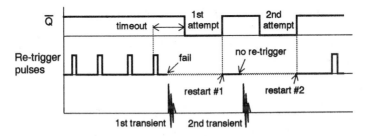

Figure 6.21 The advantage of an astable watchdog

reset after the first pulse, or more probably is derailed by another burst of interference before it can retrigger the watchdog, the watchdog will continue to bark until it achieves success (Figure 6.21). This is far more reliable than a monostable watchdog that only barks once and then shuts up.

On no account should you use a programmable timer to fulfil the watchdog function, however attractive it may be in terms of component count. It is quite possible that the transient corruption could result in the timer being programmed off, thereby completely silencing the watchdog.

Connection to the microprocessor

Figure 6.20 shows the watchdog's $\overline{Q}$ output being fed directly to the $\overline{\text{RESET}}$ input along with the power-on reset (POR) signal. In many cases it will be possible and preferable for you to trigger the timer's output from the POR signal, in order to assure a defined reset pulse width at the micro on power-up.

It is essential that you use the $\overline{\text{RESET}}$ input and not some other signal to the micro such as an interrupt, even a non-maskable one. The processor may be in any conceivable state when the watchdog barks, and it must be returned to a fully characterised state. The only state which can guarantee a proper restart is RESET.

Source of the re-trigger pulse

Equally important is that the micro should not be able to carry on kicking the watchdog when it is catatonic. At a minimum this demands AC coupling to the timer's re-trigger input, as shown by the R-C-D network in Figure 6.20. This ensures that only an edge will re-trigger the watchdog, and prevents an output which is stuck high or low from holding the timer off. The same effect is achieved with a timer whose re-trigger input is edge- rather than level-sensitive.

Using a programmable port output in conjunction with AC coupling is attractive for two reasons. It needs two separate instructions to set and clear it, making it very much less likely to be toggled by the processor executing an endless loop; this is in contrast to designs which use an address decoder to produce a pulse whenever a given address is accessed, which practice is susceptible to the processor rampaging uncontrolled through the address space. Secondly, if the programmable port device is itself corrupted but processor operation otherwise continues properly, then the retrigger pulses may cease even though the processor is attempting to write to the port. The ensuing reset will ensure that the port is fully re-initialised. Conversely, you should make sure that the port output you select is not capable of being corrupted and re-programmed to generate a square wave!

Generation of the re-trigger pulses in software

You should if possible generate the output pulse from two different software modules. The high-going edge should be generated in one module, perhaps labelled `kick_watchdog_high`, and the low-going edge in another, `kick_watchdog_low` (Figure 6.22). With a port output as described above, both edges are necessary to keep the watchdog held off. This minimises the chance of a rogue software loop generating a valid re-trigger pulse. At least one edge should only be generated at one place in the code; if a real-time "tick" interrupt is used, this could be conveniently placed at the entry to the interrupt service routine, whilst the other is placed in the background service module. This has the added advantage of guarding against the interrupt being accidentally masked off.

Placing the watchdog re-trigger pulse(s) in software is the most critical part of watchdog design and repays careful analysis. On the one hand, too many calls in different modules to the pulse generating routine will degrade the security and efficiency of the watchdog; but on the other hand, any non-trivial application software will have execution times that vary and will use different modules at different times, so that pulses will have to be generated from several different places. Two frequent critical

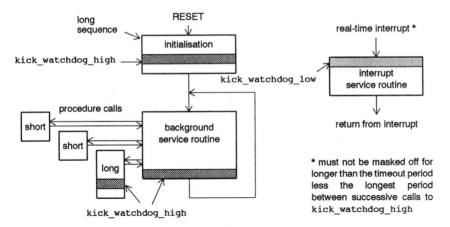

Figure 6.22 Generating the watchdog re-trigger in software

points are on initialisation, and when writing to non-volatile (EEPROM) memory. These processes may take several tens of milliseconds. Analysing the optimum placement of re-trigger pulses, and ensuring that under all correct operating conditions they are generated within the timeout period, is not a small task.

Testing the watchdog

This is not at all simple, since the whole of the rest of the circuit design is bent towards making sure the watchdog never barks. Creating artificial conditions in the software is unsatisfactory. An adequate procedure for most purposes is to subject the equipment to repeated transient pulses which are of a sufficient level to predictably corrupt the processor's operation, if necessary using specially "weakened" hardware. Be careful not to be over-enthusiastic with the spikes, or you may end up destroying several good prototypes. Build up a large enough statistical base of events to have a good chance of covering all operational conditions, and check that after each one the processor has been correctly reset and has recovered to normal operation. (This is a good task for a junior technician.) An LED on the watchdog output is useful to detect its barks. Pay particular attention to what happens when you apply a burst of spikes, so that the processor is hit again just as it is recovering from the last one. This is a vulnerable condition but unhappily it is a common occurrence in practice.

As well as testing the reliability of the watchdog, remember to include a link to disable it so that you can test new versions of software.

6.3.3   Supervisor design

The traditional method for power-on reset, and that recommended in most microprocessor applications information, is the simple R-C network across the power rail (Figure 6.23(a)). This circuit delays the voltage rise at the $\overline{\text{RESET}}$ input for a given time, long enough for the micro's required start-up period. The diode, shown dotted here, is needed to discharge the capacitor rapidly in the event of a short V_{CC} interruption. Even so, the circuit is susceptible to interruptions or dips of up to a few milliseconds, when the capacitor cannot discharge fast enough to bring the $\overline{\text{RESET}}$ input below its required threshold. It also depends on a minimum rate of rise of V_{CC}, which may not be achieved in all circumstances, and it gives no early warning of an impending power failure. This simple approach suits consumer and gadget applications, where processor unreliability is no more than mildly inconvenient.

The under-voltage detector of Figure 6.23(b) offers an improvement, whereby the RESET capacitor is held low until the input voltage to the regulator has reached a sufficiently high value for the regulator output to be stable. This point is set by R1 and R2. Any transient under-voltage will cause the comparator to rapidly discharge the capacitor and generate a reset, and the micro will also be forced to reset reliably as soon as the power fails. This still does not provide a power fail early warning, for which another comparator is needed as at Figure 6.23(c).

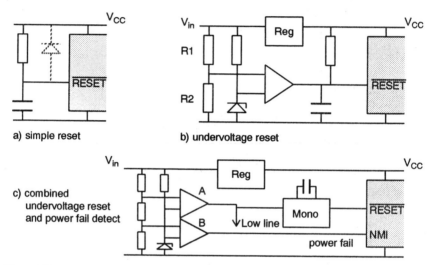

Figure 6.23 Microprocessor reset circuits

Under-voltage and power fail monitor

Comparator A switches when the minimum regulator input voltage V_{in} for a stable output is reached and serves to trigger the monostable to provide a defined-length reset pulse. Comparator B switches at a higher level of V_{in}, but one which is still below the minimum operating level. Thus when the power fails, V_{in} ramps down and first comparator B switches, then comparator A. Comparator B sets a non-maskable interrupt (NMI) at the micro which triggers the power-fail housekeeping functions. These have to be complete before comparator A switches, which triggers a reset to close down the processor completely.

A particular danger with this kind of circuit, which is effective under normal circumstances, is the threat of a "brownout". If the input voltage only droops but does not disappear altogether, the ripple on V_{in} will generate a string of power-fail pulses without activating the RESET line (Figure 6.24). Thus the software must not assume that a power-fail signal is automatically followed by a reset, and must recover smoothly from a series of power-fail interrupts of unpredictable width at the line frequency. Alternatively, the power fail signal could be buffered by a monostable as well as the reset signal.

The circuit of Figure 6.23(c) can easily be built from standard analogue ICs, although you need to check that the comparators work reliably at low voltage. The

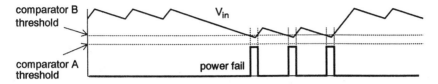

Figure 6.24 The effect of a brownout

typical LM339-type are normally quite adequate, being specified down to 2V, although their common-mode input range vanishes at this point. However, they and their associated discrete components eat up board space, and you may well prefer to use one of the purpose-designed devices that are now on the market (*cf* the comments in section 7.3.4).

Protecting non-volatile memory

Inadvertent write operations to non-volatile components – EEPROM, battery-backed CMOS RAM and real-time clocks – must be prevented in hardware when the supply rail is unstable. The behaviour of the micro's write control line is not guaranteed under these conditions, and no precautions can be taken in software. The standard technique is to gate either the write-enable or the chip-enable line to the non-volatile components, with the "low line" signal derived from comparator A in Figure 6.23(c). This ensures that no accesses are possible when low line is active. A complication with this method is that the (CMOS) gate package must be powered from the standby (battery-backed) V_{CC} rail of the non-volatile components. But the inputs of these gates are derived from the main circuit. If these inputs do not discharge fully to zero when the power is off, as may be the case with some power supply designs, then they may be unintentionally held in the CMOS gate threshold region, which can result in rapid power drain through the gate IC from the back-up power rail. ("Rapid" is a relative term – it may discharge the back-up battery in months rather than years, which you won't notice in prototype testing.) A similar effect may occur on the data/address lines to the RAM itself, though these are likely to drop to zero quickly. A pull-down (*not* pull-up) resistor on the vulnerable lines cures the problem (Figure 6.25).

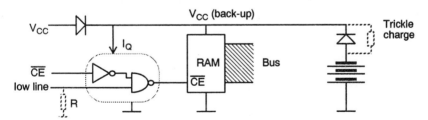

If either $\overline{CE}$ or low line are poorly defined with power-off, I_Q may be high. Pull-down R is necessary

Figure 6.25 Non-volatile memory and the back-up supply

The minimum value of this pull-down is set by the output driver's current sourcing ability. The problem with using a pull-up at this interface, or on other data bus lines, is that it provides a sneak current path from the battery supply through the bus drivers' input/output protection diodes back to the main supply, which again will discharge the back-up battery when the main supply is off (Figure 6.26).

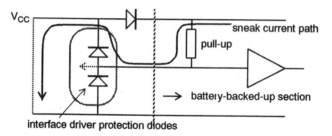

Figure 6.26 Sneak current path through pull-up resistor

V_{CC} differential

On the same subject of battery-backed RAM, another pitfall is that the RAM's V_{CC} in the circuit of Figure 6.25 is one diode drop below the operating V_{CC}. (If you minimise this by using a Schottky diode you pay the penalty of greater high-temperature reverse leakage current and hence reduced back-up time.) CMOS RAMs, particularly the early types, are in danger of latch-up or malfunction when their input signals, in this case the microprocessor address/data bus, exceed their V_{CC} pin by 0.3V. You may be able to source components which do not suffer this problem, but you must ensure that the resulting current flow which passes through their input protection diodes to the battery supply is safely limited by the output impedance of the driver. Alternatively, rather than suffer this voltage difference, you can use an active switchover, provided for instance by a MOS transistor fed from the low line signal.

A battery-back-up supply is essential if you need a real-time clock function as well as non-volatile memory. Calculating the back-up time is difficult because both battery storage capacity and more importantly, leakage current through the CMOS components and isolating devices, are highly dependent on temperature; if you have to specify a back-up time it is best only to commit yourself to a room temperature figure. If you don't need a real-time clock, use EEPROM for non-volatile memory storage and save all the extra circuitry and cost of having to incorporate a battery.

6.4 Software techniques

In a book about hardware design, it might seem strange to include a section on software. But some precautions against unwanted analogue effects can be taken in software, and it is relevant to look briefly at them here. The collection of techniques outlined below can be described as "defensive programming": recognising the possibility of data corruption and guarding against its ill effects.

6.4.1 Input data validation and averaging

Frequently, you will know in advance the range of input data that is acceptable. For instance, a thermocouple will not be producing output voltages greater than a few tens of millivolts; a thermistor or platinum resistance thermometer will not show a near-zero or negative resistance. If you can set known limits on the figures that enter as digital input to the software then you can reject data which are outside those limits.

When, as in most control or monitoring applications, each sensor inputs a continuous stream of data, this is simply a question of taking no action on false data. Since the most likely reason for false data is corruption by a noise burst or transient, subsequent data in the stream will probably be correct and nothing is lost by ignoring the bad item. Data-logging applications might require you to flag or otherwise mark the bad data rather than merely ignore it.

This technique can be extended if you have a known limit to the maximum rate-of-change of the data. You can then ignore an input which exceeds this limit even though it may be still within the range limits. It is probably due to a noise burst. Alternatively, software averaging on a stream of data to smooth out process noise fluctuations can also help remove or mitigate the effect of invalid data.

It pays to be careful when using sophisticated software for error detection, that you don't lock out genuine errors which need flagging or corrective action, such as a sensor failure. The more complex the software algorithm is, the more it needs to be tested to ensure that these abnormal conditions are properly handled.

Digital inputs

A similar checking process should be applied to digital inputs. In this case, you have only two states to check so range testing is inappropriate. Instead, given that the input ports are being polled at a sufficiently high rate, compare successive input values with each other and take no action until two or three consecutive values agree. This way, the processor will be "blind" to occasional noise glitches which may coincide with the polling time slot. This does of course mean that the polling rate must be two or three times faster than the minimum required for the specified response time, which in turn may require a faster microprocessor than originally envisaged. It is not unknown for such unanticipated noise problems to force a complete system re-design late in the project. Including the solution at the beginning will avoid this.

The same technique can be applied directly to switch contact de-bouncing, as discussed in section 6.2.2. De-bouncing in software allows you to dispense with all the extra hardware and feed a switch input directly into a digital port, as in Figure 6.14. This approach is understandably popular with keyboard users.

Interrupts

For similar reasons to those outlined above, it is preferable not to rely on edge-sensitive interrupt inputs. Such an interrupt can be set by a noise spike as readily as by its proper signal. Undoubtedly edge-sensitive interrupts are necessary in some applications, but in these cases you should treat them in the same way as clock inputs to latches or flip-flops and take extra precautions in layout and drive impedance to minimise their noise susceptibility. If you have a choice in the design implementation, then favour a level-sensitive interrupt input.

6.4.2   Data and memory protection

Volatile memory (RAM, as distinct from ROM or EEPROM) is susceptible to various forms of data corruption. These vary from "soft" errors due to cosmic radiation particles, to unintentional write accesses in severe noise environments. You cannot prevent such corruption absolutely, but you can in some cases guard against its consequences.

If you place critical data in tables in RAM, each table can then be protected by a checksum, which is stored with the table. Checksum-checking diagnostics can be run by the background routine automatically at whatever interval is deemed necessary to catch RAM corruption, and an error can be flagged or a software reset can be generated as required. The absolute values of RAM data do not need to be known provided that the checksum is recalculated every time a table is modified. Beware that the diagnostic routine is not interrupted by a genuine table modification or vice versa, or errors will start appearing from nowhere! Of course, the actual partitioning of data into tables is a critical system design decision, as it will affect the overall robustness of the system.

Data communication

The subject of data comms has a literature all to itself. The communication of digital data over long distances is prone to corruption in a statistically predictable way, and great effort has been expended to develop techniques to combat this corruption. These techniques range from simple parity checks on individual bytes through to sophisticated error detection and correction algorithms on large data blocks. This is not the place for a review of such methods. All that can be said here is that when your products use long-distance data communication, your software should incorporate some form of error detection on the received data to be at all reliable.

Unused program memory

One of the threats discussed in the section on watchdogs was the possibility of the microprocessor accessing unused memory space due to corruption of its program counter. If it does this, it will interpret whatever data it finds as a program instruction. In such circumstances it would be useful if this action had a predictable outcome.

Normally a bus access to a non-existent address returns the data #FF$_H$, provided there is a passive pull-up on the bus, as is normal practice. Nothing can be done about this. However, un-programmed ROM also returns #FF$_H$ and this can be changed. A good approach is to convert all unused #FF$_H$ locations to the processor's one-byte NOP (no operation) instruction (Figure 6.27). The last few locations in ROM can be programmed with a JMP RESET instruction, normally three bytes, which will have the effect of resetting the processor. Then, if the processor is corrupted and accesses anywhere in unused memory, it finds a string of NOP instructions and executes these (safely) until it reaches the JMP RESET, at which point it restarts.

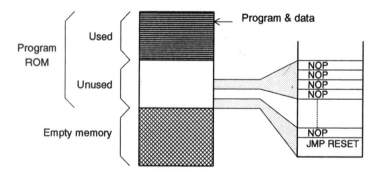

Figure 6.27 Protecting unused program memory with NOPs

The effectiveness of this technique depends on how much of the total possible memory space is filled with NOPs, since the processor can be corrupted to a random address. If the processor accesses an empty bus, its action will depend on the meaning of the #FF$_H$ instruction. The relative cheapness of large ROMs and EPROMs (64kbyte devices are now commonplace) means that you could consider using these, and filling the entire memory map with ROM, even if your program requirements are small.

6.4.3   Re-initialization

As well as RAM data, you must remember to guard against corruption of the set-up conditions of programmable devices such as I/O ports or UARTs. Many programmers seem to assume that once an internal device control register has been set up (usually in the initialization routine) it will stay that way forever. This is a dangerous assumption. Experience shows that control registers can change their contents, even though they are not directly connected to an external bus, as a result of interference. This may have consequences that are not obvious to the processor: for instance if an output port is re-programmed as an input, the processor will happily continue writing data to it oblivious of its ineffectiveness.

The safest course is to periodically re-initialize all critical registers, perhaps in the main idling routine if one exists. Timers, of course, cannot be protected in this way. The period between successive re-initializations depends on how long the software can tolerate a corrupt register, versus the software overhead associated with the re-initialization.

Chapter 7

Power supplies

The power supply is a vital but often neglected part of any electronic product. It is the interface between the noisy, variable and ill-defined power source from the outside world and the hopefully clear-cut requirements of the internal circuitry. For the purposes of this discussion it is assumed that power is taken from the conventional AC mains supply. Other supply options are possible, for instance a low-voltage DC bus, or the standard aircraft supply of 400Hz 48V. Batteries we shall discuss separately at the end of this chapter.

7.1   General

A conceptual block diagram for two common types of power supply – linear and switchmode – is given in Figure 7.1.

7.1.1   The linear supply

The component blocks of a linear supply are common to all variants, and can be described as follows:

- input circuit: conditions the input power and protects the unit, typically voltage selector, fuse, on-off switching, filter and transient suppressor

- transformer: isolates the output circuitry from the AC input, and steps down (or up) the voltage to the required operating level

- rectifier and reservoir: converts the AC transformer voltage to DC, reduces the AC ripple component of the DC and determines the output hold-up time when the input is interrupted

- regulation: stabilises the output voltage against input and load fluctuations

- supervision: protects against over-voltage and over-current on the output and signals the state of the power supply to other circuitry; often omitted on simpler circuits

7.1.2   The switch-mode supply

The advantage of the direct-off-line switch-mode supply is that it eliminates the 50Hz mains transformer and replaces it with one operating at a much higher frequency, typically 20 – 200kHz. This greatly reduces its weight and volume. The component blocks are somewhat different from a linear supply. The input circuit performs a similar function but requires more stringent filtering. This is followed immediately by a rectifier and reservoir which must work at the full line voltage, and feeds the switch element which chops the high-voltage DC at the chosen switching frequency.

The transformer performs the same function as in a linear supply but now operates

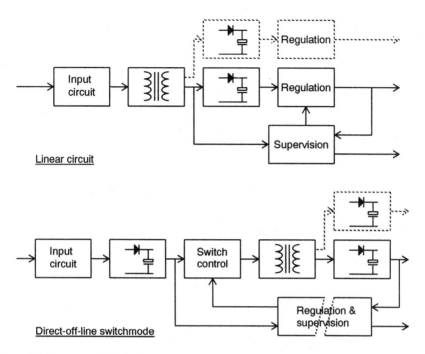

Figure 7.1 Power supply block diagrams

with a high-frequency squarewave instead of a low-frequency sinewave. The secondary output needs only a small-value reservoir capacitor because of the high frequency. Regulation can now be achieved by controlling the switch duty cycle against feedback from the output; the feedback path must be isolated so that the separation of the output circuit from the mains input is not compromised. The supervision function, where it is needed, can be combined with the regulation circuitry.

7.1.3   Specifications

The technical and commercial considerations that apply to a power supply can add up to a formidable list. Such a list might run as follows:

- input parameters: minimum and maximum voltage
  maximum allowable input current, surge and continuous
  frequency range, for AC supplies
  permissible waveform distortion and interference generation

- efficiency: output power divided by input power, over the entire range of load and line conditions

- output parameters: minimum and maximum voltage(s)
  minimum and maximum load current(s)
  maximum allowable ripple and noise
  load and line regulation
  transient response

- abnormal conditions: performance under output overload
                                     performance under transient input conditions such as
                                     spikes, surges, dips and interruptions
                                     performance on turn-on and turn-off: soft start,
                                     power-down interrupts

- mechanical parameters: size and weight
                                     thermal requirements
                                     input and output connectors
                                     screening

- safety approval requirements

- cost and availability requirements

7.1.4   Off the shelf versus roll your own

The first rule of power supply design is: do not design one yourself if you can buy it off the shelf. There are many specialist power supply manufacturers who will be only too pleased to sell you one of their standard units or, if this doesn't fit the bill, to offer you a custom version.

The advantages of using a standard unit are that it saves a considerable amount of design and testing time, the resources for which may not be available in a small company with short timescales. This advantage extends into production – you are buying a completed and tested unit. Also your supplier should be able to offer a unit which is already known to meet safety and EMC regulations, which can be a very substantial hidden bonus.

Costs

The major disadvantage will be unit cost, which will probably though not necessarily be more than the cost of an in-house designed and built power supply. The supplier must, after all, be able to make his profit. The exact economics depend very much on the eventual quantity of products that will be built; for lower volumes of a standard unit it will be cheaper to buy off the shelf, for high volumes or a custom-designed unit it may be cheaper to design your own. It may also be that a standard unit won't fulfil your requirements, though it is often worth bending the requirements by judicious circuit re-design until they match. For instance, the vast majority of standard units offer voltages of 5V (for logic) and ±12V or 15V (for analogue and interface). Life is much easier if you can design your circuit around these voltages.

A rough curve of unit costs versus power rating for a selection of readily-available standard units is shown in Figure 7.2. Typically, you can budget for £1 per watt in the 100 to 200W range. There is little cost difference between linear and switch-mode types. On the assumption that this has convinced you to roll your own, the next section will examine the specification parameters from the standpoint of design.

7.2   Input and output parameters

7.2.1   Voltage

Typically you will be designing for 240V AC in the UK, 230V in continental Europe and 115V in the US. Other countries have frustratingly minor differences. The usual supply voltage variability is ±10%, or sometimes +10%/–15%. In the UK the supply authorities are obliged to maintain their voltage at the point of connection to the

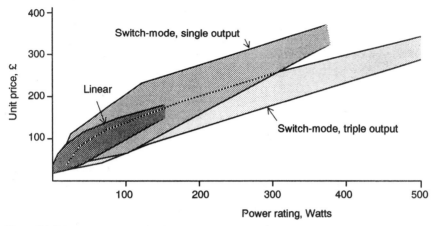

Figure 7.2 Price versus power rating for standard power supplies

customer's premises within ±6%, to which is added an allowance for local loading effects. If the voltage tolerance is applied to the UK/Europe nominal then the input voltage range becomes 207 – 264V or 195 – 264V. This range must be handled transparently by the power supply circuitry.

To cope simultaneously with both the American supply voltage, which may drop below 100V, and the European voltages is difficult for a conventional supply although it is possible to design "universal" switchmode circuits which can accept such a wide range. Traditionally, this problem has been handled by using a mains transformer with a split primary (Figure 7.3) which can be connected in series or parallel by means of a discreetly mounted voltage selector switch. This has the disadvantage that the switch

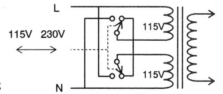

Figure 7.3 Split-primary transformer wiring

may be so discreet that the user doesn't know about it, or else it may not be discreet enough and the user may be tempted to fiddle with it. This is not a real problem in the US, but applying 240V to a unit which is set for 115V will at least annoy the user by blowing a fuse, and at worst cause real damage. Universal switch-mode supplies are therefore becoming popular.

7.2.2   Current

The maximum continuous input current should be determined by the output load and the power conversion efficiency of the circuit. The main interest in this parameter is that it determines the rating of the input circuit components, especially the protective fuse. You have to decide whether an overload on the output will open the input circuit fuse or whether other protection measures, such as output current limiting, will operate. If the input fuse must blow, you need to characterise the input current very carefully over the entire range of input voltages. It is quite possible that the difference between maximum continuous current at full load, and minimum overload current at which the fuse should blow, is less than the fusing characteristics allow. Normally you need at least a 2:1 ratio between prospective fault current and maximum operating current. This may not be possible, in which case the input fuse protects the input circuit from faults only and some extra secondary circuit protection is necessary.

7.2.3   Fuses

A brief survey of fuse characteristics is useful here. The important characteristics that are specified by fuse manufacturers are the following:

- *Rated current I_N*: that value by which the fuse is characterized for its application and which is marked on the fuse. For fuses to IEC127 (BS4265) this is the maximum value which the fuse can carry continuously without opening and without reaching too high a temperature, and is typically 60% of its minimum fusing current. For fuses to the American UL-198-G standard the rated current is 85-90% of its minimum fusing current, so that it runs hotter when carrying its rated current. The minimum fusing current is that at which the fusing element just reaches its melting temperature.

- *Time-current characteristic*: the pre-arcing time is the interval between the application of a current greater than the minimum fusing current and the instant at which an arc is initiated. This depends on the over-current to which the fuse is subjected and manufacturers will normally provide curves of the time-current characteristic, in which the fuse current is normalized to its rated current as shown in Figure 7.4. Several varieties of this characteristic are available:

  FF: very fast acting
  F: fast acting

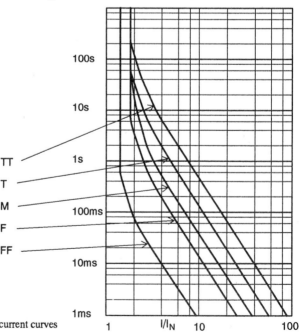

Figure 7.4 Typical fuse time-current curves

M: medium time lag
T: time lag (or anti-surge, slow-blow)
TT: long time lag

Most applications can be satisfied with either type F or type T and it is best to specify these if at all possible, since replacements are easily obtainable. Type FF is mainly used for protecting semiconductor circuits.

The total operating time of the fuse is the sum of the pre-arcing time and the time for which the arc is maintained. Normally the latter must be taken into account only when interrupting high currents, typically more than ten times the rated current.

The energy in a short-duration surge required to open the fuse depends on $I^2{\cdot}t$, and for pulse or surge applications you should consult the fuse's published I^2t rating. Current pulses that are not to open the fuse should have an I^2t value less than 50-80% of the I^2t value of the fuse.

- *Breaking capacity*: breaking capacity is the maximum current the fuse can interrupt at its rated voltage. The rated voltage of the fuse should exceed the maximum system voltage. To select the proper breaking capacity you need to know the maximum prospective fault current in the circuit to be protected – which is usually determined in mains-powered electronic products by the characteristics of the next fuse upstream in the supply. Cartridge fuses fall into one of two categories, high breaking capacity (HBC) which are sand-filled to quench the arc and have breaking capacities in the 1000's of amps, and low breaking capacity (LBC) which are unquenched and have breaking capacities of a few tens of amps or less.

7.2.4   Switch-on surge

Continuous maximum input current is usually less than the input current experienced at switch-on. An unfortunate characteristic of mains power transformers is their low impedance when power is first applied. At the instant that voltage is applied to the primary, the current through it is limited only by the source resistance, primary winding resistance and the leakage inductance. A separate component of this current is the abnormal secondary load due to the low impedance of the un-charged power supply reservoir capacitor.

The effect is most noticeable on modern toroidal mains transformers when the mains voltage is applied at its peak halfway through the cycle, as in Figure 7.5. The typical mains supply has an extremely low source impedance, so that the only current limiting is provided by the transformer primary resistance and by the fuse. Toroidals are particularly efficient and can be wound with relatively few turns, so that their series resistance and leakage inductance is low; the surge current can be more than ten times the operating current of the transformer.[†] In these circumstances, the fuse usually loses out. The actual value of surge depends on where in the cycle the switch is closed, which is random; if it is near the zero crossing the surge is small or non-existent, so it is possible for the problem to pass unnoticed if it is not thoroughly tested.

Several solutions are possible. One is to specify an anti-surge or time-lag (type T or TT) fuse. This will rupture at around twice its rated current if sustained for tens or hundreds of seconds, but will carry a short overload of ten or twenty times rated current

---

† The effect happens with all transformers, but is more of a problem with toroidals.

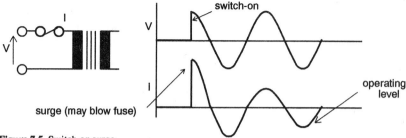

Figure 7.5 Switch-on surge

for a few milliseconds. Even so, it is not always easy to size the fuse so that it provides adequate protection without eventually failing in normal use, particularly with the high ratios of surge to operating current that can occur. A re-settable thermal circuit breaker is sometimes more attractive than a fuse, especially as it is inherently insensitive to switch-on surges.

Current limiting

A more elegant solution is to use a negative-temperature-coefficient (NTC) thermistor in series with the transformer primary and fuse. The device has a high initial resistance which limits the inrush current but in so doing dissipates power, which heats it up. As it heats, its resistance drops to a point at which the power dissipated is just sufficient to maintain the low resistance and most of the applied voltage is developed across the transformer. The heating takes one or two seconds during which the primary current increases gradually rather than instantaneously.

NTC thermistors characterised especially for use as inrush current limiters are available, and can be used also for switchmode power supply inputs, motor soft-start and filament lamp applications. Although the concept of an automatic current-limiter is attractive, there are three major disadvantages:

- because the devices operate on temperature rise they are difficult to apply over a wide ambient temperature range;
- they run at a high temperature during normal operation, so require ventilation and must be kept away from other heat-sensitive components;
- they have a long cool-down period of several tens of seconds and so do not provide good protection against a short supply interruption.

PTC thermistor limiting

Another solution to the inrush current problem is to use instead a *positive*-temperature-coefficient thermistor in place of the fuse. These are characterised such that provided the current remains below a given value self-heating is negligible, and the resistance of the device is low. When the current exceeds this value under fault conditions the thermistor starts to self-heat significantly and its resistance increases until the current drops to a low value. Note that such a device does not protect against electric shock and so cannot replace a fuse in all applications, but because of its inherent insensitivity to surges it can be useful in local protection of a transformer winding.

A further solution, which is usually too complex for anything but specialised applications, is to switch the ac input voltage only at the instant of zero crossing, using a triac. This results in a predictable switch-on characteristic, and may be attractive if electronic switching is required for other reasons. Inrush current is also a potential problem in direct-off-line switchmode supplies, where the reservoir capacitor is charged directly through the mains rectifier, and comparatively complex "soft-start" circuits may be needed in order to protect the input components.

7.2.5   Waveform distortion

An increasing problem for electricity supply systems is the proportion of semiconductor-based equipment in the supply load. This is because the load current that such equipment takes is pulsed rather than sinusoidal. Current is only drawn at the peak of the input voltage, in order to charge the reservoir capacitors in the power supply. The normal rms-to-average ratio of 1.11 for a sinusoidal current is considerably higher for this type of waveform (Figure 7.6).

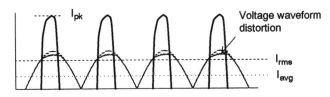

Figure 7.6 Peak input current in a rectifier/ reservoir power supply

The ratio of the peak load current I_{pk} to I_{rms} is called the "crest factor" and here it depends on the input impedance of the reservoir circuit. The lower the impedance, the faster the reservoir capacitor(s) will charge, which results in lower output voltage ripple but higher peak current.

The significance of crest factor is that it affects the power handling capability of the supply network. A network of a given sinusoidal rms current rating will show considerable extra losses when faced with loads of a high crest factor. The supply mains does not have zero impedance, and the result of the extra network voltage drop at each crest is a waveform distortion in which the sinusoidal peak is flattened. This is a form of harmonic distortion and its seriousness depends on the susceptibility of other loads and components in the network.

Peak current summation

Large systems installations, in which there are many electronic power supplies of fairly high rating fed from the same supply, are the main threat. The current peaks always occur together and so reinforce each other. A network which is dominated by resistive loads such as heating and lighting, as in domestic premises, can tolerate a proportion of high crest factor loads more easily.

The relevant standard for harmonic generation of household appliances and similar equipment is IEC555-2 (equivalent to EN60555-2 or BS5406 Part 2 in the UK), and it is likely that legislation will appear in the future to make this standard more widely applied than it is at present. This requirement will find its way down to the power supply designer, and you may at some point be faced with the need either to increase the impedance of the reservoir circuit, or incorporate extra components at the input for crest factor correction.

Interference

Electrical interference generated within equipment and conducted out through the mains supply port is already subject to regulation for some product sectors in some countries, and with the adoption of the European Community EMC Directive it will become mandatory for *all* electrical or electronic products to comply with interference limits. The usual method of reducing such interference is to use a radio frequency filter at the mains supply inlet. Switch-mode power supplies are usually the worst offenders, because they generate large interference currents at harmonics of the switching frequency well into the hf region. The size and weight advantages of switch-mode supplies may be increasingly balanced by the need to fit larger filters as the interference limits become tougher.

Chapter 8 covers aspects of mains input filtering in greater detail.

7.2.6   Frequency

The UK and European mains frequency is held to 50Hz ±1%. The American supply standard is 60Hz. The difference in frequencies does not generally cause any problem for equipment that has to operate off either supply (provided that it's designed in Europe!), since mains transformers and reservoir circuits that perform correctly at 50Hz will have no difficulty at 60Hz. The sensitivity of the power supply circuits to supply voltage droops at 60Hz should be less than at 50Hz since the ripple amplitude is only 83% of the 50Hz figure, and the minimum voltage will thus be higher (Figure 7.7).

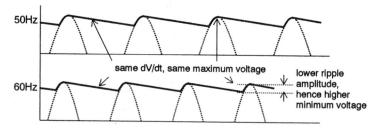

Figure 7.7 Ripple voltage versus frequency

The ±1% tolerance on the mains frequency is slightly misleading because the supply authorities maintain a long-term tolerance very much better than this. Diurnal variations are arranged to cancel out, and this allows the mains to be used as a timing source for clocks and other purposes. If you are planning to use the mains frequency for internal timing then you will need to incorporate some kind of switching arrangement if the equipment will be used on both US and European systems.

7.2.7   Efficiency

The efficiency of a power supply module is its output power divided by its input power. The difference between the two quantities is accounted for by power losses in the various components in the power supply.

$$\text{Efficiency } \eta \;=\; P_{out}/P_{in} \;=\; P_{out}/(P_{out} + P_{loss})$$

The efficiency normally worsens as the load is reduced, because the various losses and quiescent operating currents assume a greater proportion of the input power. Therefore, if you are concerned about efficiency, do not use a power supply that is heavily over-rated for its purpose. Linear supply efficiency also varies considerably with its input voltage, being worst at high voltages, because the excess must be lost across the regulator. Switch-mode supplies do not have this problem.

Normally efficiency is not of prime concern for mains power supplies, since it is not essential to make optimum use of the available power. At higher powers the heat generated by an inefficient unit can be troublesome. It is far more important that a power converter for a portable instrument should be efficient because this directly affects useable battery life.

Linear power supplies are rarely more than 50% efficient unless they can be matched to a narrow input voltage range, whereas switch-mode supplies can easily exceed 70% and with careful design can reach 85%. For this reason switch-mode supplies are more popular, despite their greater complexity, at the higher power levels and for battery-powered units.

Sources of power loss

The components in a power supply which make the major contribution to losses are

- the transformer: core losses, determined by the operating level and core material, and copper losses, determined by I^2R where R is the winding resistance

- the rectifiers: diode forward voltage drop, V_F, multiplied by operating current; more significant at low output voltages

- linear regulator: the voltage dropped across the series pass element multiplied by the operating current; greatest at high input voltages

- switching regulator: power dissipated in the switching element due to saturation voltage, plus switching losses in this and in snubber and suppressor components, proportional to switching frequency

If you sum the approximate contribution of each of these factors you can generally make a reasonable forecast of the efficiency of a given power supply design. The actual figure can be confirmed by measurement and if it is wildly astray then you should be looking for the cause.

7.2.8   Deriving the input voltage from the output

In a linear supply with a series pass regulator element, the design must proceed from the minimum acceptable output voltage at maximum load current and minimum input voltage. These are the worst-case conditions and determine the input voltage step-down required. The minimum DC input voltage is given by the minimum output voltage plus all the tolerances and voltage drops in series:

$$V_{in,dc} = V_{out(min)} + V_{tol,reg} + V_{series,reg} + V_{series, CS} \cdots$$

where $V_{out(min)}$ is the minimum acceptable output voltage
$V_{tol,reg}$ is the regulator voltage tolerance, assuming it is not adjustable
$V_{series,reg}$ is the voltage drop across the regulator series pass element
$V_{series, CS}$ is the voltage drop across the current sense element if fitted

All the above parameters are specified at full load current. This value for $V_{in,dc}$ is then the minimum input voltage allowed for a dc-input supply, or it is the voltage at the minimum of the ripple trough for a rectified and smoothed ac-input supply. This is related to the transformer secondary voltage as follows:

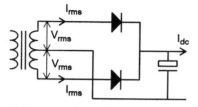

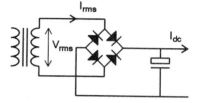

Full-wave centre tap: $I_{rms} \approx 1.2 \cdot I_{dc}$ Full-wave bridge: $I_{rms} \approx 1.8 \cdot I_{dc}$

Figure 7.8 Rectifier configuration

$$V_{tx} = (V_{in,dc} + V_{ripple} + V_D)/0.92 \cdot (V_{ac(nom)}/V_{ac(min)}) \cdot 1/\sqrt{2}$$

where V_{ripple} is the peak ripple voltage across the reservoir capacitor
V_D is the voltage drop across the rectifier diode(s)
V_{tx} is the rms transformer secondary voltage
$V_{ac(nom)}$ is the specified transformer input voltage
$V_{ac(min)}$ is the minimum line input voltage
All parameters at full load current

The figure of 0.92 is an approximate allowance for full-wave rectifier efficiency with a single-capacitor reservoir. It can be more accurately derived using curves published by Schade [†]. Complications set in because the current drawn through the secondary is not sinusoidal, but occurs at the crest of the waveform (see section 7.2.5). The extra peak current reduces the peak secondary voltage from its quoted value, if this value is specified for a resistive load. You can get around this either by knowing the transformer's losses in advance and allowing for the extra IR drop, or by specifying the transformer for a given circuit and letting the transformer supplier do the work for you, if you're buying a custom component. The transformer secondary rms current rating is determined by the rectifier configuration (Figure 7.8).

Take as an example a typical linear regulator circuit supplying 5V ±5% at 1A.

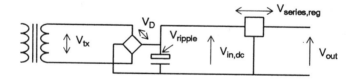

Here, $V_{out(min)}$ is allowed to be 5V – 5% = 4.75V. The regulator we shall use is a standard 7805 type with ±4% tolerance and so $V_{tol,reg}$ is 5V · 0.04 = 0.2V. Its specified minimum series voltage drop (or dropout voltage) at 1A and a junction temperature of 25°C (note the temperature restriction) is 2.5V maximum. The required minimum input voltage is

$$V_{in,dc} = 4.75 + 0.2 + 2.5 = \underline{7.45V}$$

If the peak ripple voltage is 2V and each diode forward drop in the bridge is 1V, then the transformer voltage with a 240V nominal spec but a minimum line voltage of 195V will need to be

$$V_{tx} = [7.45 + 2 + (2 \times 1)] / 0.92 \cdot 240/195 \cdot 1/\sqrt{2} \qquad = \underline{10.83Vrms}$$

---

† O.H.Schade, **Analysis of Rectifier Operation**, Proc.IRE, vol 31 1943, pp 341-361

From this example you can see that the secondary-side input voltage needed to assure a given output voltage is very much higher than the actual output voltage. One of the major culprits is the dropout voltage of the regulator which in this example accounts for at least 50% of the output power, although it becomes proportionally less at higher output voltages. Low-dropout voltage regulators which use a pnp transistor as the series pass element, such as National Semiconductor's LM2930 range, are popular for this reason and also where the minimum input voltage can be close to the output level, as in automotive applications.

Power losses at high input voltage

You can also see more clearly in the above example where the power losses are which contribute to reduced efficiency. When the input voltage is increased to its maximum value the dissipation in the series-pass element is worst. In the above example with the mains input at 264V, the average value of $V_{in,dc}$ rises to 12.5V, and 7.45V of this must be lost across the regulator, which because it is passing the full load current amounts to one-and-a-half times the load power! The advantage of the switch-mode supply is that it adjusts to varying input voltages by modifying its switching duty cycle, so that an increased input voltage automatically reduces the input current and the overall power taken by the unit remains roughly constant.

7.2.9   Low-load condition

When the output load is removed or substantially reduced then the dissipation in the power supply will drop. This is good news for almost all parts of the circuit, except for the voltage rating of the components around $V_{in,dc}$. When there is a combination of low load and maximum supply input voltage, the peak value of $V_{in,dc}$ is highest. A crucial factor here is the transformer regulation. This is the ratio

$$\text{Regulation} = (V_{sec,unloaded} - V_{sec,loaded})/V_{sec,loaded}$$

and a small or poorly-designed transformer can have a regulation exceeding 20%. If this figure is used for the transformer in the above example then the peak V_{tx} off-load at maximum input voltage will rise to 20.2V. At the same time the diode forward voltage drops at low current will be much less, say 0.6V each, so the possible maximum voltage at the reservoir capacitor could be around 19V. Thus even the common 16V rated electrolytic will not be adequate for this circuit. For higher voltage outputs, the maximum input voltage can even exceed the voltage rating of the regulator itself, and you have to invest in a pre-regulator to hold the maximum to a manageable level. Note that this condition is not the worst-case for regulator power dissipation, because the regulator is not passing significant load current.

Maximum regulator dissipation

In fact maximum series-pass dissipation does not necessarily occur at full load current, because as the current rises the voltage across the series-pass element falls. The maximum dissipation will occur at less than full output if the voltage dropped across the dc supply's equivalent series resistance is greater than half the difference between the no-load input voltage and the output voltage. Figure 7.9 shows this graphically.

Minimum load requirement

A further problem, particularly with switch-mode supplies, is that the stability of the regulator cannot always be assured down to zero load. For this reason some rails have

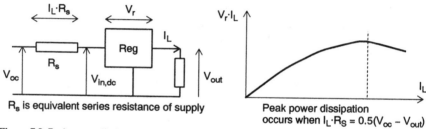

Figure 7.9 Peak power dissipation

to be run with a minimum load, such as a bleed resistor, to remain within specification, and this represents an unnecessary additional power drain. Many circuits, of course, always take a minimum current and so the minimum load is not then a problem.

7.2.10   Rectifier and capacitor selection

The specification of the rectifiers and capacitors is dominated by surge and ripple current concerns.

Reservoir capacitor

The minimum capacitor value is easily decided from the required ripple voltage:

$$C \quad = \quad I_L / V_{ripple} \cdot t$$

where I_L is the dc load current
V_{ripple} is the acceptable ripple voltage
For mains inputs, t is about 2ms less than the ac input period, 8ms for 50Hz or 6ms for 60Hz full-wave

A more accurate value can be derived from Schade's curves[†] which have been reprinted in numerous textbooks, but remember that the tolerance on reservoir capacitors is wide (typically ±20%) and accuracy is rarely needed.

For load currents exceeding 1A, ripple current rating (see page 81) tends to determine capacitor selection rather than ripple voltage. As is made clear throughout this chapter, the peak current flow through the rectifier/capacitor circuit is many times higher than the dc current, due to the short time in each cycle for which the capacitor is charging. The rms ripple current is 2 - 3 times higher than the dc load. Ripple current rating is directly related to temperature and you may need to derate the component further if you need high ambient temperature and/or high reliability operation.

As an example, a load current of 2A and a permissible ripple voltage of 3V at 100Hz suggests a 5300µF capacitor. Typical capacitors of the next value up from this, 6800µF, have 85°C ripple current ratings from 2 to 4A. The higher ratings are larger and more expensive. But actual ripple current requirements will be 4 – 6A. To meet this you will need to use either a much larger capacitor (typically 22,000µF), or two smaller capacitors in parallel, or derate the operating temperature *and* use a slightly larger capacitor. If you don't do this, your design will become yet another statistic to prove that electrolytic capacitors are the prime cause of power supply failure.

Rectifiers

Although in the full-wave arrangements (Figure 7.8) the diodes only conduct on

† O.H.Schade, **Analysis of Rectifier Operation**, Proc.IRE, vol 31 1943, pp 341-361

alternate half cycles, because the rms current is 2 - 3 times higher than the dc load current a rating of *at least* the full load current, and preferably twice it, is necessary. Surge current on turn-on may be much higher, especially in the higher power supplies where the ratio of reservoir capacitance to operating current is increased. This is of even greater concern in direct-off-line switch-mode supplies where there is no transformer series resistance to limit the surge, and a diode rating of 2 - 5 times the average dc current is needed.

The maximum instantaneous surge current is V_{max}/R_s and the capacitor charges with a time constant of $\tau = C \cdot R_s$, where R_s is the circuit series resistance. As a conservative guide, the surge won't damage the diode if τ is less than a half-cycle at mains frequency and V_{max}/R_s is less than the diode's rated I_{FSM}. All diode manufacturers publish I_{FSM} ratings for a given time constant; for example, the typical 1N5400 series with 3A average rating have an I_{FSM} of 200A. You may discover that you have to incorporate a small extra series resistance to limit the surge current, or use a larger diode, or apply the techniques discussed on page 206.

The rectifier's peak-inverse-voltage (PIV) rating needs to be at least equivalent to the peak ac voltage for the full-wave bridge circuit, or twice this for the full-wave centre tap. But you should increase this considerably (by 50 to 100%) to allow for line transients. This is easy for low-voltage circuits, since 200V diodes cost hardly any more than 50V ones, and does not normally make much cost difference in mains circuits. For 240V, a minimum of 600V PIV and preferably 800V PIV should be specified, even if you are using a transient suppressor at the input.

7.2.11   Load and line regulation

Load regulation refers to the permissible shift in output voltage when the load is varied, usually from none to full. Line (or input) regulation similarly refers to the permissible shift in output voltage when the input is varied, usually from maximum to minimum. Provided that the design of the input circuit has been properly considered as described above, so that the input voltage never goes outside the regulator's operational range, these parameters should be wholly a function of the regulator circuit itself. The regulator is essentially a feedback circuit which compares its output voltage against a reference voltage, so the regulation depends on two parameters: the stability of the reference, and the gain of the feedback error amplifier. If you use a monolithic regulator IC, then these factors are taken into account by the manufacturer who will specify regulation as a data sheet parameter.

Thermal regulation

A monolithic regulator IC includes the voltage reference on-chip, along with other circuitry and the series pass element. This means that the reference is subject to a thermal shift when the power dissipation of the series pass element changes. This gives rise to a separate longer term component of regulation, called thermal regulation, defined as the change in output voltage caused by a change in dissipated power for a specified time. Provided the chip has been well-designed, thermal regulation is not a significant factor for most purposes, but it is rarely specified in data sheets and for some precision applications may render monolithic regulators unsuitable.

Load sensing

No three-terminal regulator can maintain a constant voltage at anywhere other than its output terminals. It is common in larger systems for the load to be located at some distance from the power supply module, so that load-dependent voltage drops occur in the wiring connecting the load to the power supply output (*cf* section 1.1.5 on page 8). This directly impacts the achievable load regulation.

The accepted way to overcome this problem is to split the regulator feedback path, and incorporate two extra "sensing" terminals which are connected so as to sense the output voltage at the load itself (Figure 7.10). The voltage drop across this extra pair of wires is negligible because they only carry the signal current. The voltage at the regulator output is adjusted so as to regulate the voltage at the sensing terminals.

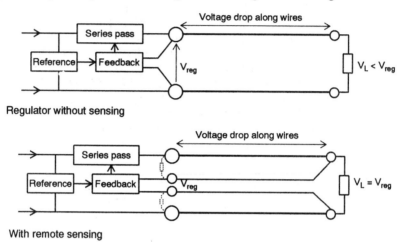

Figure 7.10 Load sensing

The minimum voltage at the regulator input must be increased to allow for the extra output voltage drop. It is wise to connect coupling resistors (shown dotted in Figure 7.10) from the output to sense terminals, so as to ensure correct operation when the sense terminals are accidentally or deliberately disconnected. Sensing can only offer remote load regulation at one point and so is not really suited when one power supply module feeds several loads at different points.

7.2.12   Ripple and noise

Ripple is the component of the AC supply frequency (or more often its second harmonic) which is present on the output voltage; noise is all other AC contamination on the output. In a linear power supply, ripple is the predominant factor and is given by the AC across the reservoir capacitor reduced by the ripple rejection (typically 70–80dB) of the regulator circuit. A figure of less than 1mV rms should be easy to obtain. HF noise is filtered by the reservoir and output capacitors and there are no significant internal noise sources, provided that the regulator isn't allowed to oscillate, so that apart from supply-frequency ripple linear power supplies are very "quiet" units.

Switching noise

The same cannot be said for switch-mode power supplies. Here the noise is mainly due to output voltage spikes at the switching frequency, caused by fast-rise-time edges and HF ringing at these edges feeding through, or past, filtering components to the output. The ESR and ESL of typical output filter capacitors (see page 81) limits their ability to attenuate these spikes, while the self-inductance of ground wiring limits the high frequency effectiveness of ground decoupling anyway. Switch-mode output ripple and noise is typically 1% of the rail voltage, or 100 – 200mV. In fact comparing ripple and noise specifications is the easiest way to distinguish a linear from a switch-mode unit, if there is no other obvious indication. The bandwidth over which the specification applies is important, since there is significant energy in the high-order harmonics of the switching noise, and at least 10MHz is needed to get a true picture. Because of stray coupling over this extended bandwidth the noise frequently appears in common-mode, on both supply and 0V simultaneously, and is then very difficult to control. Differential-mode noise spikes can be reduced dramatically by including a ferrite bead in series, and a small ceramic capacitor in parallel with the output capacitor.

The presence of switching noise is not a problem for digital circuits, but it creates difficulties for sensitive analogue circuits if their bandwidth exceeds the switching frequency. It can cause interference on video signals, mis-clocking in pulse circuits and voltage shifts in DC amplifiers. These effects have to be treated as EMC phenomena (see chapter 8) and can be cured by suitable layout, filtering and shielding, but if you have the option in the early stages to choose a linear supply instead, take it – you will save yourself a lot of trouble.

Layout to avoid ripple

Power supply output ripple is aggravated by incorrect layout of the wiring around the reservoir capacitor. This is a specific instance of the common-impedance interference coupling that was discussed in chapter 1.

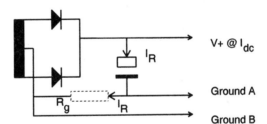

Figure 7.11 Incorrect
reservoir connection

At first sight grounds A and B in Figure 7.11 look equivalent. But there will be a potential between them of $I_R \cdot R_g$, where I_R is the capacitor ripple current and R_g is the track or wiring resistance common to the two grounds through which the ripple current flows. (The ripple current path is through the transformer, the two diodes and the capacitor.) This current is only drawn on peaks of the ac input waveform to charge the reservoir capacitor, and its magnitude is only limited by the combined series resistance of the transformer winding, the diodes, capacitor and track or wiring. If the steady-state dc current supplied is 1A then the peak ripple current may be of the order of 5A; thus 10mΩ of R_g will give a peak difference of 50mV between grounds A and B. If some parts of the circuit are grounded to A and some to B, then tens of millivolts of hum injection are included in the design at no additional cost, and increasing the reservoir value to try and reduce it will actually make matters worse as the peak ripple current is increased. You can check the problem easily, by observing the output ripple on a scope; if it has a pulse shape then wiring is the problem, if it looks more like a sawtooth then you need more smoothing.

Correct reservoir connection

The solution to this problem, and the correct design approach, is to ground *all* parts of the supplied circuit on the supply side of the reservoir capacitor, so that the ripple current ground path is not common to any other part of the circuit (Figure 7.12). The same applies to the V+ supply itself. The common impedance path is now reduced to the capacitor's own ESR, which is the best you can do.

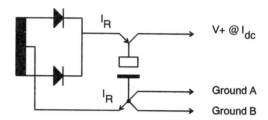

Figure 7.12 Correct
reservoir connection

7.2.13   Transient response

The transient response of a power supply is a measure of how fast it reacts to a sudden change in load current. This is primarily a function of the bandwidth of the regulator's feedback loop. The regulator has to maintain a constant output in the face of load changes, and the speed at which it can do this is set by its frequency response as with any conventional operational amplifier. The trade-off that the power supply designer

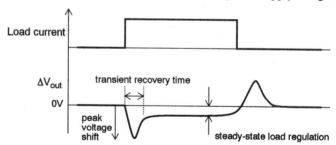

Figure 7.13 Load
transient response

has to worry about is against the stability of the regulator under all load conditions; a regulator with a very fast response is likely to be unstable under some conditions of load, and so its bandwidth is "slugged" by a compensation capacitor within the regulator circuit. Too much of this and the transient response suffers. The same effect can be had by siting a large capacitor at the regulator output, but this is a brute-force and inefficient approach because its effect is heavily load-dependent. Note that the 78XX series of three-terminal regulators should have a small, typically 0.1μF capacitor at the output for good transient response and HF noise decoupling. This is separate from the required 0.33 – 1μF capacitor at the input to ensure stability.

Switch-mode vs. linear

The transient response of a switch-mode power supply is noticeably worse than that of a linear because the bandwidth of the feedback loop has to be considerably less than the switching frequency. Typically, switch-mode transient recovery time is measured in milliseconds while linear is in the tens of microseconds.

If your circuit only presents slowly-varying loads then the power supply's transient response will not interest you. It becomes important when a large proportion of the load can be instantaneously switched – a relay coil or bank of LEDs for example – and the rest of the load is susceptible to short-duration over- or under-voltages.

Although load transient response is usually the most significant, a regulator also exhibits a delayed response to line transients, and this may become important when you are feeding it from a DC input which can change quickly. The line transient response is normally of the same order as the load response.

7.3   Abnormal conditions

7.3.1   Output overload

At some point in its life, a power supply is almost sure to be faced with an overload on its output. This can take the form of a direct short circuit across its output due to the slip of a technician's screwdriver, or a reduced load resistance due to component failure in the load circuit, or incorrect connection of too many loads. It may also be mistaken connection to the output of another power supply. The overload can be transient or sustained, and at the very least any power supply should be designed to withstand a continuous short circuit at its output(s) without damage. This is almost universally achieved with one of two techniques: constant current limiting or foldback current limiting.

Constant current limiting

Output overloads threaten mainly the series pass element in a linear supply, or the switching element in a switch-mode supply. In either case, an output over-current will subject the device to the maximum current that the input can supply while it is sustaining the full input-output differential voltage, and this will put its dissipation well outside its safe operating area (SOA) boundary (see Figure 4.7). Swift destruction will ensue.

Constant current limiting operates by ensuring that the output current available from the power supply limits at a maximum that is only marginally above the full load rating of the unit. Figure 7.14 shows this operation for a linear supply. This simple

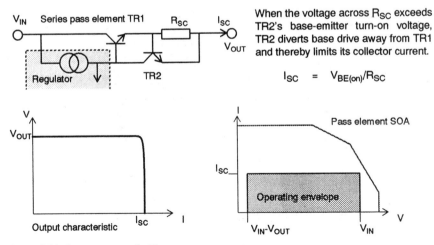

When the voltage across R_{SC} exceeds TR2's base-emitter turn-on voltage, TR2 diverts base drive away from TR1 and thereby limits its collector current.

$$I_{SC} = V_{BE(on)}/R_{SC}$$

Figure 7.14 Constant current limiting

circuit works quite well but the actual value of I_{SC} is very dependent on TR2's V_{BE}, and hence on temperature, so that either you must allow a large margin over full load current or use a more complex circuit.

Switch-mode current limiting is more complex yet because you need to limit on a cycle-by-cycle basis to fully protect the switching element, so that current sensing on the output line is insufficient. Several techniques have been evolved to achieve this; consult switching regulator design manuals for details.

Foldback current limiting

A disadvantage of constant current limiting is that to obtain sufficient SOA the pass element must have a much higher collector current capability than is needed for normal operation. "Foldback" current limiting reduces the short circuit current whilst still allowing full output current during normal regulator operation, thereby giving more efficient use of the pass element's SOA.

The development of the constant-current circuit to give foldback operation is shown in Figure 7.15. Although foldback allows the use of a smaller series pass element, it has its limitations. As the foldback ratio, I_K/I_{SC}, is increased, the required value of R_{SC} increases and this calls for a greater input voltage at high foldback ratios. There is an absolute limit to the foldback ratio when R_{SC} is infinite of

$$[I_K/I_{SC}]_{max} = 1 + (V_{OUT}/V_{BE(on)})$$

and so foldback ratios of greater than 2 or 3 are impractical for low voltage regulators.

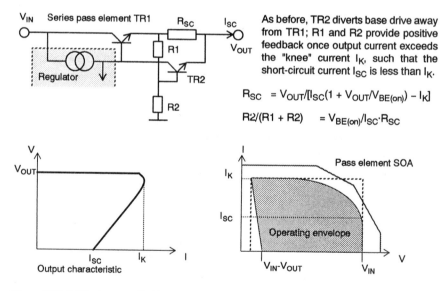

As before, TR2 diverts base drive away from TR1; R1 and R2 provide positive feedback once output current exceeds the "knee" current I_K, such that the short-circuit current I_{SC} is less than I_K.

$$R_{SC} = V_{OUT}/[I_{SC}(1 + V_{OUT}/V_{BE(on)}) - I_K]$$

$$R2/(R1 + R2) = V_{BE(on)}/I_{SC} \cdot R_{SC}$$

Figure 7.15 Foldback current limiting

7.3.2 Input transients

Under this heading we need to consider spikes, surges and interruptions on the input supply.

Interruptions

On the mains supply, dips ("brownouts") and outages of up to 500ms are fairly common, due to surge currents and fault clearing in the supply network. Other sources of supply may also experience such interruptions. The occurrence of longer supply breaks depends very much on location. In the UK, the average consumer loses power for 90 minutes in the year, but a rural consumer on the end of a long overhead line may experience much longer interruptions, while an urban consumer with several redundant supply routes may see none at all.

A power supply should be able to cope with short interruptions and brownouts transparently, so that the load is unaffected by them. The "hold-up time" (Figure 7.16) specifies for how long the output remains stable after loss of input, and it can be

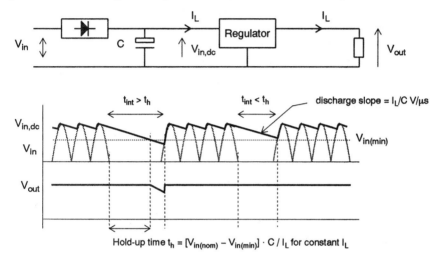

Hold-up time $t_h = [V_{in(nom)} - V_{in(min)}] \cdot C / I_L$ for constant I_L

Figure 7.16 Hold-up time

anywhere from a few to several hundred milliseconds. It is almost entirely determined by the size of the main reservoir capacitor, since this provides the only source of power when the input is removed. A linear regulator can be considered as a constant-current sink discharging this capacitor and therefore it is easy to calculate the hold-up time for a given load and input voltage. A switch-mode regulator draws more current as its input voltage drops, so accurately determining hold-up time for this type requires the solution of a current-time integral. The higher the operating voltage, the easier it is to obtain a long hold-up time, because energy storage in the reservoir is proportional to $0.5 \cdot C \cdot V^2$. This gives another advantage to direct-off-line switching supplies, whose main reservoir operates at the full line voltage.

Taking the quoted parameters for the linear supply of page 210, what values does this give for its hold-up time at full load at 240V and 204V?

The ripple on $V_{in,dc}$, of 2V at 1A, with a full-wave rectified supply so that its period is 10ms, means that the reservoir capacitor is

$$C = I \cdot t / V = 1 \cdot 10.10^{-3} / 2 = 5000\mu F$$

At 240V the minimum value for $V_{in,dc}$ at the ripple trough is

$$V_{in,dc(min)} = 14.05 - 2_{(ripple)} - 2_{(diode)} \qquad = 10.05V$$

so the hold up time at this voltage given a minimum requirement of 7.45V at the regulator is

$$t_h = (10.05 - 7.45) \cdot 5000.10^{-6} / 1 \qquad = \underline{13ms}$$

At 204V (240V − 15%) the minimum value for $V_{in,dc}$ is 7.94V, so the hold-up time is now

$$t_h = (7.94 - 7.45) \cdot 5000.10^{-6} / 1 \qquad = \underline{2.5ms}$$

It is clear that hold-up time specified at nominal input voltage may be considerably less when the power supply is running at its minimum input voltage. In fact, the minimum input voltage as calculated in section 7.2.8 is that for which the hold-up time is zero. All this is assuming the worst-case condition, that the supply is interrupted at the minimum of the ripple trough. If hold-up time is important for your circuit, you must decide at what input voltage it is to be specified.

Spikes and surges

Chapter 8 discusses the occurrence of transient overvoltages on mains and automotive supplies. Some precautions need to be taken to prevent these as far as possible from propagating through the power supply and impacting the load circuit. Short, low-energy but fast rise-time transients can only be dealt with by good circuit layout, minimising ground inductance and stray coupling, and by input filtering. Slower but higher-energy transients call for the use of transient suppressor devices at various points in the power supply, and for over-voltage protection.

7.3.3   Transient suppressors

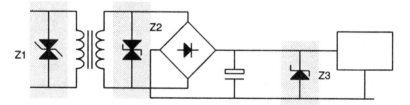

Figure 7.17 Transient suppression in a linear supply

Figure 7.17 shows three positions for transient suppressors in a linear supply. The advantages and disadvantages of each position can be summarized as follows:

- Z1: protects all components in the unit from differential-mode surges but is subject to the lowest source impedance. This means that it must have a high energy rating to withstand the maximum expected surge without destruction, and it will have a fairly high ratio of clamped voltage to normal running voltage. In effect, voltage surges up to about twice the peak operating voltage will be let through.

- Z2: this is a more effective position as it still protects the vulnerable rectifiers, but is itself protected by the additional source impedance of the transformer. It can therefore be a smaller component but still have a good ratio of clamped to peak operating voltage. It has no effect on spikes which may have been converted from differential to common mode by the interwinding capacitance of the transformer.

- Z3: this protects the regulator and subsequent circuitry but not the rectifiers. It is something of a "belt-and-braces" position, but it does suppress input common-mode spikes that the previous positions would have let through. It should be sized so that its peak clamping voltage is just less than the absolute maximum input voltage of the regulator. Smaller surges then rely on the transient response of the regulator to contain them.

7.3.4   Overvoltage protection

If the circuit that your power supply is driving is very expensive and susceptible to overvoltages – for instance it may include a £100 microprocessor which must not be subject to more than 7V – then it is worth including extra circuitry at the power supply output for overvoltage protection. The first time that it operates, it will have saved the extra expense of designing it in.

This might be as simple as a 6.2 or 6.8V zener diode across the output of a 5V supply. See page 114 for a discussion of how to size such a zener. This does not offer foolproof protection, because if the overvoltage is sustained and derives from a low source impedance – perhaps the series-pass element has failed – then the zener is likely to fail itself, and may fail open-circuit, in which case it has been wasted. Something more drastic is called for, and the conventional solution is a crowbar.

This gets its name from the method of ensuring that no voltage is present between two live terminals, by the simple expedient of putting a crowbar – which is assumed to be able to carry any likely short-circuit current – across them. In its more sophisticated version in electronic power supplies, the crowbar takes the form of a triggered thyristor. The thyristor is permanently in place across the output, or in some designs across the reservoir, but it is only triggered when a supervisory circuit detetcts the presence of an overvoltage. It then stays triggered, holding the output voltage to V_H, until the current

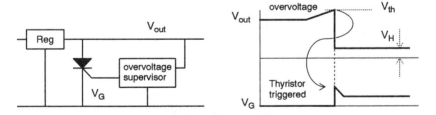

Figure 7.18 Overvoltage protection

through it is interrupted. Although this current may be high, the voltage across it is not, so its dissipation is fairly low. Obviously the power supply itself must be protected against a sustained output short circuit, either by current limiting or a fuse or preferably both. Figure 7.18 shows the operating principle.

Crowbar circuit requirements

The thyristor must be capable of dumping, virtually instantaneously, both the continuous short-circuit current of the supply and the energy stored in the reservoir capacitors. It must therefore have a high single-pulse I^2t and di/dt rating. Some manufacturers are characterising devices especially for this purpose, and the di/dt performance is helped by making sure the trigger pulse has a fast edge and is well in excess of the minimum gate current requirement.

Both the supervisory circuit and the thyristor itself must be immune from false triggering due to short transients, as the nuisance value of an unnecessary shutdown may exceed that of a real overvoltage in some instances. Some degree of delay in the trigger pulse is essential, and characterising the overall system (power supply plus crowbar protection plus load) for the acceptable and necessary delay and overvoltage threshold is the most critical part of overvoltage protection design.

7.3.5   Turn-on and turn-off

Sometimes, the behaviour of the power rails when the input power is applied or removed is important to the load circuit. The power rail never instantaneously reaches its operating level as soon as the input is applied. It will ramp up to the rail voltage as the reservoir and other capacitors charge, and it may overshoot its nominal voltage briefly if the regulator frequency compensation has not been optimised – this is a particular danger with some switch-mode circuits. It may suffer from noise or oscillation due to the switch-on process as it ramps up. Particularly if the load circuit includes a microprocessor, it will not be safe to start the circuit operation until the rail voltage has settled. You may require the power supply to have a flag output which signals to the load circuit that all is well. This output is often connected to the micro's RESET input. (This is discussed from the micro's point of view in section 6.3.3.)

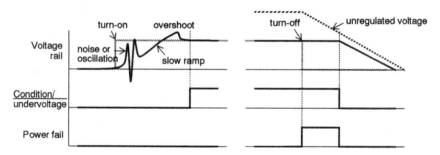

Figure 7.18 Power rail supervision waveforms

Similarly, when the power is switched off, the microprocessor needs to be able to power down in an orderly fashion. This is best achieved by generating a power-fail interrupt as soon as a power failure is detected, followed by an under-voltage warning when the power rail starts to droop. The time delay between the two will be roughly equivalent to the hold-up time as discussed earlier, and this delay must be long enough to enable the micro to perform its power-down housekeeping functions. Required outputs are shown in Figure 7.18.

PSU supervisor circuits

All the functions of undervoltage and power fail monitoring, and overvoltage protection, can be gathered up into a single power supply supervisory circuit, and several ICs are on the market for this purpose. Examples are the MC3524, SG3543, ICL7665 and 7677, TL7705 and MAX691. These chips are basically a collection of comparators and delay circuits, integrated into one package for ease of use. Unfortunately, there are a multiplicity of types and as yet few second sources, and the parts cost may be greater than you would suffer when using standard comparators such as the LM339. In many cases you may still prefer to design your own supervisory circuit from such standard components.

A typical application will require the supervisory circuit to have inputs from the dc output rail for overvoltage protection, the reservoir capacitor for undervoltage warning, and the low-voltage AC input for power fail detection. Outputs will go to the crowbar device and the load circuit. Bear in mind that the supervisor needs to operate reliably down to very low supply voltages.

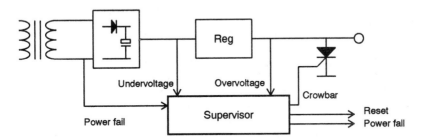

Figure 7.19 Configuration of power supply supervisor

7.4   Mechanical requirements

7.4.1   Case size and construction

If you are designing a supply as an integral part of the rest of the equipment then generally you won't need to consider its mechanical characteristics separately from the equipment design. If you are buying in a standard unit, or designing yourself a modular unit which will be used for different products, then case construction becomes important. Standard products tend to fall into one of four categories:

- open frame, chassis mounting
- enclosed, chassis mounting
- encapsulated, pcb or chassis mounting
- rack module

Both linear and switch-mode types are available in all these variants, but power rating, connections and the need for screening play an important role in the final selection.

Open frame

This is normally the cheapest option, since all that is supplied is a pcb mounted on a simple metal chassis which serves as a rudimentary heatsink. Connections are made by wiring to terminals or spade lugs mounted on the board. No environmental protection or screening is offered and the power unit must be enclosed completely within the equipment it is supplying. Open frame units are most popular in the 10 to 100W range, with models available up to 250W.

Enclosed

Cased power supplies are more common for power ratings above 100W. They offer reasonably effective screening which is important for switch-mode supplies, and can incorporate a fan for efficient convective cooling, which is not possible with open or encapsulated types but is necessary at high powers. The greater electronics cost tends to mask the cost of the extra mechanical components. Connections are made to screw terminals on the outside of the case, and the internal circuitry is guarded from wandering fingers and other foreign bodies.

Encapsulated

Fully encapsulated units are available up to 40W, and can be either pcb-mounted via pins or chassis-mounted using screw terminals for the connections. Their great advantage is that they can be treated as just another component during equipment production, and do not need any further environmental protection for their internal circuit. EMI screening can be incorporated as part of the encapsulating box. Higher power ratings than 40W require a heatsink outside the encapsulation. The encapsulation tends to provoke reliability problems when much heat has to be dissipated, and if you are going for a higher-power unit it would be wise to seek concrete reliability data. Encapsulation is particularly popular for low-power dc-to-dc converters which can be incorporated within a system at board level, to generate different and/or regulated supplies from a common dc bus.

Rack mounting modules or cassettes

With the increasing popularity of rack-mounted modular processor equipment, usually based on the Eurocard rack and DIN-41612 connector standard, there is a corresponding need for power supply modules that can share the same rack. These are available from 25 up to 500W. All but the smallest are switch-mode types, since space and thermal capacity are strictly limited, and applications are mainly digital. Connection is by mating plug and socket, mounted on the card frame, and it is vital to ensure that the connector used is capable of carrying the load current without loss, and is rated for mains voltages. The DIN-41612 H15 connector is widely used, with a leading earth pin to maintain safety when withdrawing or inserting the unit.

7.4.2   Heatsinking

A necessary requirement for the continued health of any semiconductor device, be it monolithic IC regulator, rectifier diode or power transistor, is that its junction temperature should stay within safe limits. Junction temperature is directly related to power dissipation, thermal resistance and ambient temperature, and the function of a heatsink is to provide the minimum acceptable thermal resistance between the junction and its environment.

Chapter 9 includes a discussion of thermal management, including methods of calculating the size requirements of heatsinks. Here it is enough to say that the power supply often represents the most concentrated source of heat in an item of equipment. As soon as its efficiency is roughly known, you should calculate the heat output and take steps to ensure that the mechanical arrangement will be allow an efficient heat flow. At the minimum this will involve ensuring that all components which will need heatsinking are positioned to allow this, and that the power unit's positioning within the overall equipment gives adequate thermal conductivity to the environment. Too many designs end up with a fan tacked onto the case as an afterthought!

7.4.3   Safety approvals

The major safety risk for power supplies is the threat of electric shock due to contact with "live parts". Safety is discussed in greater depth in section 9.1. One of the important but forgotten functions of a power supply is to ensure a safe segregation of the low-voltage circuitry, which may be accessible to the user, from the high-voltage input, which must be inaccessible. Segregation is normally assured in a power supply by maintaining a minimum distance around all parts that are connected to the mains, including spacing between the primary and secondary of the transformer. This, of course, adds extra space to the design requirements.

There are many national authorities concerned with setting safety requirements. Foremost among these are UL in the United States, CSA in Canada, and VDE in Germany. In the UK, there are various British Standards for different product sectors. As designer, you can either choose to apply a particular set of requirements for your company's market, or if you plan to export worldwide, you can discover the most stringent requirements and apply these across the board. Frequently this turns out to be the German standard VDE0806, which calls for 3750V isolation withstand voltage between primary and secondary circuits. A less rigorous but more common standard is 2500V isolation. You will find most standrad power supplies quote one or other of these levels. If no safety specification is quoted, beware.

It may not be legally necessary to have your product approved to safety regulations, but it is often commercially desirable. Using a bought-in supply which already has safety approval goes a long way to helping your own equipment achieve it. Note that there is a difference, on data sheets, between the words "designed to meet..." and "certified to..." The former means that, when you go for your own safety approval, the approvals agency will still want to satisfy themselves, at your expense, that the power supply does indeed meet their requirements. The latter means that this part of the approvals procedure can be bypassed. It therefore puts the unit cost of the power supply up, but saves you some part of your own approval expenses.

7.5   Batteries

Battery power is mainly used for portability or stand-by (float) purposes. All batteries operate on one or another variant of the principle of electro-chemical reaction, in which anode (negative) and cathode (positive) terminals are separated by an electrolyte, which is the vehicle for the reaction. This basic arrangement forms a "cell", and a battery consists of one or more cells. The chemistry of the materials involved is such that a potential is developed between the electrodes which is capable of sustaining a discharge current. The voltage output of a particular cell type is a complex function of time, temperature, discharge history and state of charge.

The basic division is between primary (non-rechargeable) and secondary (rechargeable) cells. This section will survey the various types of each shortly, but first we shall make a few general observations on designing with batteries.

7.5.1   Initial considerations

When you know you are going to use a battery, select the cell type as early as possible in the circuit and mechanical design. This allows you to take the battery's properties into account and increases the likelihood of a cost-effective result, as otherwise you will probably need a larger, or more expensive, battery or will suffer a reduced equipment specification. Having made the selection, you can then design the circuit so that it works over the widest possible part of the battery's available voltage range. Some of the cheaper types deliver useful power over quite a wide range, and some of this energy will be lost if the design cannot cope with it. Also, check that the battery can deliver the circuit's load current requirements over the working temperature. This capability varies considerably for different chemical systems.

Always aim to use standard types if your specification calls for the user to be able to replace the battery. Not only are they cheaper and better documented, but they are widely distributed and are likely to remain so for many years. You should only need to use special batteries if your environmental conditions or energy density requirements are extreme, in which case you have to make special provisions for replacement or else consider the equipment as a throwaway item.

Voltage and capacity ratings

Different types of battery have different nominal open-circuit voltages, and the actual terminal voltage falls as the stored energy is used. Manufacturers provide a discharge characteristic curve for each type which indicates the behaviour of voltage against time for given discharge conditions. Note that the open-circuit voltage can exceed the voltage under load by up to 15%, and the operating voltage is normally significantly less than the nominal battery voltage.

The capacity of a battery is expressed in ampere-hours (Ah) or milliampere-hours (mAh). It may also be expressed in normalised form as the "C" figure, which is the nominal capacity at a given discharge rate. This is more frequently applied to rechargeable types. Capacity will be less than the C rating if the battery is discharged at a faster rate; for instance, a 15Ah lead-acid type discharged at 15 amps (1C) will only last for about 20 minutes (Figure 7.22).

Series and parallel connection

Cells can be connected in series to boost voltage output, but doing so decreases the reliability of the overall battery and there is a risk of the weakest cell being driven into reverse voltage at the end of its life. This increases the likelihood of leakage or rupture, and is the reason why manufacturers recommend that all cells should be replaced at the same time. Good design practice minimises the number of series-connected cells. There are now several ICs which can be used to efficiently multiply the voltage output of even a single cell. It is not that difficult to design a switching converter that simultaneously boosts and regulates the battery voltage.

Parallel connection can be used to increase the capacity or discharge capability, or the reliability of the battery. Increased reliability requires a series diode in each parallel path to isolate failed cells. Some battery systems are hazardous if a cell is inserted or connected the wrong way round (they can explode) and if the user can do it, he or she will, possible exposing your company to a product liability claim. It is therefore best to restrict parallel connection to specially-assembled units.

On the same subject, reverse insertion of the whole battery will threaten your circuit, and if it is possible, the user will do it. Either incorporate assured polarity into the battery compartment or provide reverse polarity protection, such as a fuse or series diode, at the equipment power input.

Mechanical design

Choose the battery contact material with care to avoid corrosion in the presence of moisture. The recommended materials for primary cells are nickel-plated steel, austenitic stainless steel or inconel, but definitely not copper or its alloys. The contacts should by springy in order to take up the dimensional tolerances between cells. Single-point contacts are adequate for low current loads, but you should consider multiple contacts for higher current loads. The simplest solution is to use ready-made battery compartments or holders, provided that they are properly matched to the types of cell you are using. PCB-mounting batteries have to be hand soldered in place after the rest of the board has been built, and you need to liaise well with the production department if you are going to specify these types.

Rechargeable batteries when under charge, and all types when under overload, have a tendency to out-gas. Always allow for safe venting of any gas, and since some gases will be flammable, don't position a battery near to any sparking or hot components. If severe vibration or shock is part of the environment, remember that batteries are heavy and will probably need extra anchorage and shock absorption material.

Storage, shelf life and disposal

Maximum shelf life is obtained if batteries are stored within a fairly restricted temperature and humidity range. Self discharge rate invariably increases with temperature. Different chemical systems have varying requirements, but extreme temperature cycling should be avoided, and you should arrange for tight stock control with proper rotation of incoming and outgoing units, to ensure that an excessively aged battery is not used. Rechargeable types should be given a regular top-up charge. Because they may contain toxic heavy metals such as cadmium and mercury, there are restrictions on disposal of some types, and you need to consult the manufacturer on this point. Indeed, there is legislation in some countries to reduce the mercury content of batteries, and in a few years' time you may be forced towards using environmentally-sound battery systems. Don't burn used batteries unless you have an explosives licence!

7.5.2   Primary cells

The main chemical systems employed in primary, non-rechargeable cells are zinc carbon, alkaline manganese, mercury, silver oxide, zinc air and lithium. Figure 7.21 compares the typical discharge characteristics for these types on medium load.

Zinc carbon

This until fairly recently was the cheap-and-cheerful type you could get at any newsagents and chemists. Its low price was matched by its performance. It has now been largely superseded, even for consumer applications, by the alkaline manganese type, and is not recommended for new designs.

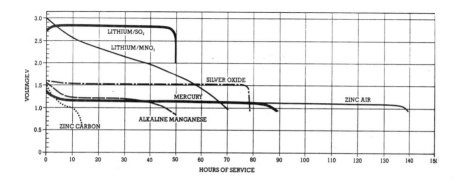

Figure 7.21 Medium-load discharge characteristics for various primary cells
Source: Duracell

Alkaline manganese

The operating voltage range of this type, which uses a highly conductive aqueous solution of potassium hydroxide as its electrolyte, is 1.3 to 0.8V per cell under normal load conditions, while its nominal voltage is 1.5V. Recommended end voltage is 0.8V per cell for up to 6 series cells at room temperature, increasing to 0.9V when more cells are used. The alkaline battery is well suited to high-current discharge. It can operate between −30 and +80°C, but high relative humidity can cause external corrosion and should be avoided. Shelf life is good, typically 85% of stored energy being retained after 3 years at 20°C. Standard types are now widely and cheaply available in retail outlets and it can therefore be confidently used in most general-purpose applications.

Mercury

This system is based on the zinc/mercuric oxide couple and its advantages are a stable operating voltage and high energy density. It is preferred for applications where voltage stability or space are at a premium, and is popular in "button" cell form for digital watches and similar small items. Its effective voltage range is typically 1.3 to 1.0V per cell, with a recommended end voltage of 0.9V. It has similar environmental and storage life properties to alkaline manganese, but is not so good at sub-zero temperatures. It is not efficient at discharge rates higher than the 20-hour rate.

Silver oxide

Zinc-silver oxide cells are used as button cells with similar dimensions and energy density to the mercury types, but they are five or six times as expensive for a given capacity. Their advantage is that they offer a higher operating voltage, typically 1.5V, which is stable for some time and then decays gradually, and can provide intermittent high pulse discharge rates and better low temperature operation. They are popular for such applications as photographic equipment. Typical shelf life is two years at room temperature.

Zinc air

This type has the highest volumetric energy density, but is very specialised and not widely available. It is activated by atmospheric oxygen and can be stored in the sealed state for several years, but once the seal is broken it should be used within 2 months. It has a comparatively narrow environmental temperature and relative humidity range. Consequently its applications are somewhat limited. Its open circuit voltage is typically 1.45V, with the majority of its output delivered between 1.3 and 1.1V. It cannot give sustained high output currents.

Lithium

Several battery systems are available based on the lithium anode with various electrolyte and cathode compounds. Lithium is the lightest known metal and the most electro-negative element. Their common features are a high terminal voltage, very high energy density, wide operating temperature range, very low self discharge and high cost. They have been used for military applications for some years. If abused, they can be potentially very hazardous, which causes less concern in military circles than it does for consumers, and they are not widely available commercially for replacement purposes. There are restrictions on air transport of lithium batteries. Certain types are now becoming popular for single-cell memory backup and other low current drain applications, because of their high voltage and "fit-and-forget" lifetime characteristics. Operating voltages range from 2.5 to 3.5V. Very high pulse discharge rates (up to 30A)

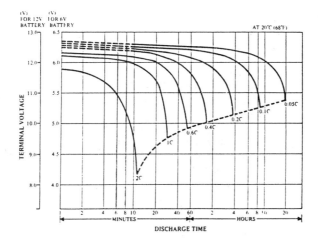

Figure 7.22 Discharge characteristics for sealed lead-acid batteries
Source: Yuasa

are possible.

7.5.3   Secondary cells

In contrast to the various primary systems, there are only two common rechargeable types: lead-acid and nickel-cadmium. These have quite different characteristics. Neither of them offer anywhere near the energy density of primary cells.

Lead-acid

The lead-acid battery is the type which is known and loved by millions all over the world, especially on cold mornings when it fails to start the car. As well as the conventional automotive version, it is widely available in a sealed or "maintenance-free" variant in which the sulphuric acid electrolyte is retained in a glass mat and does not need topping-up. This version is of more interest to circuit designers as it is frequently used as the standby battery in mains-powered systems which must survive a mains failure.

These types have a nominal voltage of 2V, a typical open circuit voltage of 2.15V and an end-of-cycle voltage of 1.75V per cell. They are commonly available in standard case sizes of 6V or 12V nominal voltage, with capacities from 1 to 100Ah. Typical discharge characteristics are as shown in Figure 7.22. The symbol "C", as noted earlier, is the ampere-hour rating, conventionally quoted at the 20-hour discharge rate (5-hour discharge rate for nickel-cadmium). Ambient temperature range is typically from –30 to +50°C, though capacity is reduced to around 60%, and achievable discharge rate suffers, at the lower extreme.

Sealed lead-acid types can be stored for a matter of months at temperatures up to 40°C, but will be damaged, perhaps irreversibly, if they are allowed to spend any length of time fully discharged. Self discharge is quite high and increases with temperature. You will therefore need to ensure that a recharging regime is followed for batteries in stock. For the same reason, equipment which uses these batteries should only have them fitted at the last moment, preferably when it is being despatched to the customer or on installation.

Typical operational lifetime in standby float service, if proper float charging is followed, is four to five years. When the battery is frequently discharged a number of factors affect its service life, including temperature, discharge rates and depth of discharge. A battery discharged repeatedly to 100% of its capacity will have only perhaps 15% of the cyclic service life of one that is discharged to 30% of its capacity. Overrating a battery for this type of duty has distinct advantages.

Nickel-cadmium

NiCads, as they are universally known, are comparable in energy density and weight to their lead-acid competitors but address the lower end of the capacity range. Typically they are available from 0.15 to 7 Ah. Nominal cell voltage is 1.2V, with an open circuit voltage of 1.4V and an end-of-cycle voltage of 1.0V per cell. This makes them comparable to alkaline manganese types in voltage characteristics, and you can buy nicads in the standard cell sizes from several sources, so that your equipment can work off primary or secondary battery power.

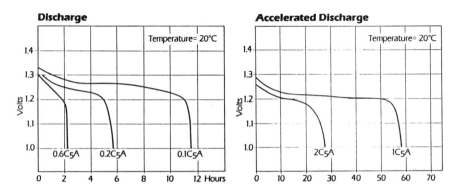

Figure 7.23 NiCad discharge characteristics
Source: Saft

NiCads offer an ambient temperature range from –40 to +50°C. They are well suited to trickle charging and are widely used for memory back-up purposes; batteries of two, three or four cells are available with pcb mounting terminals which can be continuously trickle charged from the logic supply, and can instantly supply a lower back-up voltage when this supply fails. Self-discharge rate is high and a cell which is not trickle charged will only retain its charge for a few months at most. Unlike lead-acid types they are not damaged by long periods of full discharge, and because of their low internal resistance they can offer high discharge rates.

7.5.4   Charging

The re-charging procedures are quite different for the two types of secondary cell, and you will drastically shorten their lifetime if you follow the wrong one. The greatest danger is of over-charging. Briefly, NiCads require constant current charging, while lead acid needs constant voltage.

Lead-acid

The recommended method for these types is to provide a current-limited constant voltage (Figure 7.24). The initial charging current is limited to a set fraction of the C value, generally between 0.1 and 0.25. The constant voltage is set to 2.25 – 2.5V per cell, depending on whether the intention is to trickle charge or recover from a cyclic discharge. The actual voltage is mildly temperature dependent and should be compensated by -4mV/°C when operating with extreme variations. This charging characteristic can easily be provided by a current-limited voltage regulator IC.

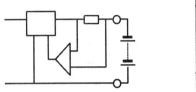 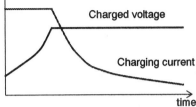

Figure 7.24 Constant-voltage charging

An elegant modification to this circuit is to arrange for the output voltage to drop from the cyclic charge level to the trickle charge level when the charging current has decreased sufficiently, typically to 0.05C. This two-step charging gives the advantages of rapid recovery from deep discharge along with the benefit of trickle charging without threat to the battery's life. It does not work too well when the load circuit is connected, though.

The cheap-and-cheerful charging method is to charge the battery from a full-wave or half-wave rectified AC supply through a series resistance. This is known as "taper" charging because the applied voltage rises towards the constant voltage level as the charging current tapers away. Since it is cheap it is very popular for automotive use, but manufacturers do not recommend it because of the risk of over-charging. Fluctuations in the mains supply can easily lead to over-voltage and the end current cannot be properly controlled. If you have to use it, include a timer to limit the overall charging time.

Lead-acid batteries *can* be charged from a constant current, typically between 0.05 and 0.2C, subject to monitoring the cell voltage to detect full charge. The technique is not often used, but can be effective for charging several batteries in series at once.

NiCad

Because the voltage characteristic during charging varies substantially and actually drops when the cell is over-charged, NiCads can only be charged from a constant current. Continuous charging at up to the 0.1C rate is permissible without damage to the cell. A 0.1C charge rate will recharge the cell in 16 hours, not 10, due to the inefficiency of the charging process. An accelerated charge rate of up to 0.3C is permissible for long periods without harm, but the cell temperature will rise when charging is complete. Higher rates for a rapid charge will work, but it is essential to monitor the charging progress and terminate it before the battery overheats.

A series resistor from a voltage source at a significantly higher level than the battery terminal voltage, is an adequate constant-current charger. For tighter control over the charging current, and especially for rapid charging, a current regulator with a battery temperature sensor is needed.

Chapter 8

Electromagnetic Compatibility

8.1   The need for EMC

All electrical and electronic devices generate electromagnetic interference, and are susceptible to it. It is your job as product designer to reduce this generation and susceptibility to acceptable levels. With the increasing penetration of solid-state electronics into all areas of activity, acceptable levels of interference have become progressively tighter as physical separation between devices has reduced and reliance on their operation has increased. Solid-state, particularly integrated circuit technology, is more susceptible than the vacuum-tube devices of years ago, and the popularity of plastic cases with their lack of screening is a further factor. The ability of a device to operate within the limits of interference immunity and suppression is known as electromagnetic compatibility (EMC).

In some areas of electronics EMC has been a product requirement for a long time. Military electronics has severe limitations imposed on it, often because of the proximity of high-power pulse equipment (radars) to sensitive signal processing equipment in the same aircraft, ship or vehicle, and military EMC standards first appeared in the 1960s. The increasing use of walkie-talkies on process plant and elsewhere has prompted users of safety-critical instrumentation to specify a minimum immunity from rf interference. Measuring and weighing equipment must be prevented from giving incorrect readings in the presence of interference, and domestic broadcast receivers should be able to work alongside home computers.

Radio frequencies are not the only source of interference. Transients can be generated by power switching circuits, lightning, electric motors, spark ignition devices or electrostatic discharge, to name just a few. Microprocessor circuits are particularly susceptible to impulse interference and must be protected accordingly.

The importance of EMC

Many manufacturers have found out the hard way that poor EMC performance of a product can be extremely costly, both in terms of damaged reputation and in the measures needed to improve performance once a fault has been found. For this reason, many firms will test their products for EMC before releasing them even though there may be no applicable standard for that class of equipment. Others, whose markets may be less demanding or whose products are not used in a crowded electromagnetic environment, have not yet considered it necessary to specify EMC performance.

This situation is now changing, both because of electromagnetic pollution which is bringing to light more cases of poor EMC, and because impending legislation will compel manufacturers to consider their products' EMC. We shall look at the legal side shortly. The technical constraints on equipment EMC are that the equipment should continue to function reliably in a hostile electromagnetic environment, and that it

Immunity	Emissions
• mains voltage drop-outs, dips, surges and distortion • transients and radio frequency interference (RFI) conducted into the equipment via the mains supply • radiated transient or RFI, picked up and conducted into the equipment via signal leads • RFI picked up directly by the equipment circuitry • electrostatic discharge	• mains distortion, transients or RFI generated within the equipment and conducted out via the mains supply • transient or RFI, generated within the equipment and conducted out via signal leads • RFI radiated directly from the equipment circuitry, enclosure and cables

Table 8.1 Electromagnetic compatibility effects

should not itself degrade that environment to the extent of causing unreliable operation in other equipment. EMC therefore splits neatly into two areas, labelled "immunity" and "emissions", and a list of some of the interference types and coupling methods is shown in Table 8.1.

8.1.1 Immunity

The electromagnetic environment within which equipment will operate and which will determine its required immunity can vary widely, as can the permissible definition of reliable operation. For instance, the magnitude of radiated fields encountered depends critically on the distance from the source. Strong RF fields occur in the vicinity of radars, broadcast transmitters and rf heating equipment (including microwave ovens at 2.45GHz). The electric field strength falls off linearly with distance from the transmitting antenna, provided that your measuring point is in the far field, defined as greater than $\lambda/2\pi$ where λ is the wavelength. The value of the field strength in volts per metre can be calculated from

$$E = (30 \cdot P)^{0.5} / d$$

where   P is the radiated power in watts, or power fed to the antenna times the antenna gain
d is the distance from the antenna in metres

In the near field closer than $\lambda/2\pi$ the field strength can be much greater, but it depends on the type of antenna and how it is driven.

Radio transmitters

AM broadcast-band transmitter power levels tend to be around 100 – 500kW, but the transmitters are usually located away from centres of population. Field strength levels in the 1 – 10V/m range may be experienced occasionally, but at medium frequencies coupling to circuit components is inefficient, so these transmitters do not pose a great problem. TV and FM-band transmitters are more often found close to office or industrial environments and the greatest threat is to equipment on the upper floors of tall buildings, which may be in line-of-sight to a nearby transmitter of typically 10kW. Field strengths greater than 10V/m are possible, and although the building structure may give some attenuation, achieving immunity from even a 1V/m field is not trivial at these frequencies, where cable and track lengths approach resonance and coupling is correspondingly more efficient.

Portable transmitters (walkie-talkies, cellphones) do not have a high radiated power but they can be brought extremely close to susceptible equipment. Typical field strengths from a 1W uhf hand-held transmitter are 5 – 7 V/m at half a metre distance.

Radars

Another serious threat is from radars in the 1 – 10GHz range, particularly around airports. It is not difficult to measure a pulsed 50V/m field strength up to 3km from these radars. Again, building attenuation may give some relief, but set against this is the problem that pulsed RFI is particularly upsetting to microprocessor circuits. As an aside, civil aircraft regulatory authorities have published a defined "RF environment"[†] in terms of field strength versus frequency that civil aircraft may be expected to meet, and should therefore be protected against. The worst-case threat is from certain US ground-based radars in the 2 – 4GHz range, where it is assumed that the aircraft can fly close to the antenna through the main beam, and the peak field strength is defined to be 17kV/m (no, that is not a misprint, *seventeen thousand* volts per metre).

From these considerations a minimum of 3V/m from, say, 10MHz to 1GHz represents a reasonable design criterion for RFI immunity, with 10V/m being preferred. Immunity from pulsed interference above 1GHz is extremely hard to quantify.

Transients

Conducted transient immunity is becoming relatively more important, because microprocessor-based circuits are much more susceptible to transient upset than are analogue circuits. Mains transients, from many sources, are far more common than is generally realised. A study by the German ZVEI[‡] made a statistical survey of 28,000 live-to-earth transients exceeding 100V, at 40 locations over a total measuring time of about 3,400 hours. Their results were analyzed for peak amplitude, rate of rise and energy content. Table 8.2 shows the average rate of occurrence of transients for four classes of location, and Figure 8.1 shows the relative number of transients as a function of maximum transient amplitude. This shows that the number of transients varies roughly in inverse proportion to the cube of peak voltage.

Rate of rise was found to increase roughly in proportion to the square root of peak voltage, being typically 3V/ns for 200V pulses and 10V/ns for 2kV pulses. Other field experience has shown that mechanical switching usually produces multiple transients (bursts) with risetimes as short as a few nanoseconds and peak amplitudes of several hundred volts.

As a general guide, microprocessor equipment should be tested to withstand pulses at least up to 2kV peak amplitude. Thresholds below 1kV will give unacceptably

---

† Users's Guide, **Protection of aircraft electrical and electronic systems against the effects of the external radio frequency environment,** EUROCAE WG33 Subgroups 2 & 3

‡ See **Transients in Low Voltage Supply Networks,** J.J.Goedbloed, IEEE Transactions on Electromagnetic Compatibility, Vol EMC-29 No 2, May 1987, pp 104-115

Area Class	Average rate of occurrence (transients/hour)
Industrial	17.5
Business	2.8
Domestic	0.6
Laboratory	2.3

Table 8.2 Average rate of occurrence of mains transients

Source:
Transients in Low Voltage Supply Networks, J.J.Goedbloed, IEEE Transactions on Electromagnetic Compatibility, Vol EMC-29 No 2, May 1987, p 107

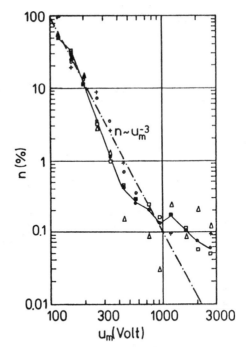

Figure 8.1 Relative number of transients vs. maximum transient amplitude
● Total, ☐ Industrial, Δ Business, ○ Domestic + Lab

frequent corruptions in nearly all environments, while between 1kV – 2kV occasional corruption will occur. For a belt-and-braces approach for high reliability equipment, a 4 – 6kV threshold is not too much.

Other sources of conducted transients are telecommunication lines and the automotive 12V supply. The automotive transient environment is particularly severe with respect to its nominal supply range. The most serious automotive transients (Figure 8.2) are the load dump, which occurs when the alternator load is suddenly disconnected during heavy charging; switching of inductive loads, such as motors and solenoids; and alternator field decay, which generates a negative voltage spike when the ignition switch is turned off.

ESD

Electrostatic discharge is a further source of transient upset which most often occurs when a person who has been charged to a high potential by movement across an insulating surface then touches an earthed piece of equipment, thereby discharging themselves through the equipment. The achievable potential depends on relative humidity and the presence of synthetic materials (Figure 8.3), and the human body is roughly equivalent to a 150pF capacitance in series with 150Ω resistance, so that currents of tens of amps can flow for a short period with a very fast (sub-nanosecond) rise-time. Even though it may have low energy and be conducted to ground through the equipment case, such a current pulse couples very easily into the internal circuitry.

Determining and specifying the effects of interference

Required reliability of operation depends very much on the application. Entertainment

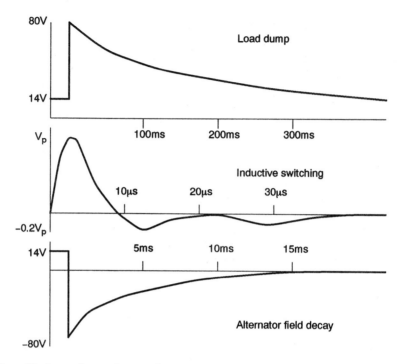

Figure 8.2 Automotive transient waveforms

devices and gadgets come fairly close to the bottom of the scale, whereas computers and instrumentation for control of critical systems such as fly-by-wire aircraft and nuclear power systems come near the top. The purchasers and authorities responsible for such systems have recognised this for some years and there is a well-established raft of EMC requirements for them, but in less critical sectors this is not the case and EMC performance is mainly determined by market factors.

Actually determining whether a piece of equipment has been affected by interference is not always easy. Interference may cause a degradation in accuracy of measuring equipment, it may cause noticeable deterioration in audio quality or it may

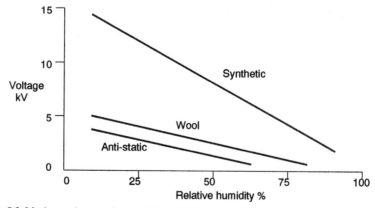

Figure 8.3 Maximum electrostatic charging voltages
Source: IEC 801 Part 2

corrupt a microprocessor program. If the processor circuit has a program-recovery mechanism (see section 6.3.2) this may correct the corruption before it is apparent, or the effects of the interference may be confused with a software glitch. Determining cause and effect in the operating environment is particularly difficult when the interference is transient or occasional in nature. When you are contemplating immunity testing, it is essential to be sure what will constitute acceptable performance and what is a test failure.

8.1.2   Emissions

By comparison with immunity requirements, equipment emissions are relatively easy to characterise. The majority of emissions from electronic equipment are due to either switching or other electromechanical operations, or digital clock lines. The former can be pulses at mains frequency such as from thyristor phase controllers, motor commutator noise, individual switching "clicks" or switch-mode power supply harmonics, and they are generally conducted out of the equipment via the mains lead. These emissions have been regulated for many years in order to minimise interference to AM broadcast and communications services.

Emissions from digital equipment

Digital equipment with high-frequency square wave clocks generates noise into the hundreds of MHz. The system clocks and their harmonics are the principal source because their energy is concentrated into a narrow band, but wideband noise from the data and address lines is also present. The noise amplitude and spectral distribution may vary depending on the operating mode of the circuit or its resident software. Emissions from some classes of digital equipment, such as personal computers, have been regulated by some countries, again to control interference to broadcast and communication services.

Equipment emissions can be either conducted or radiated. Commercial emission standards differentiate between the two on the basis of frequency; the breakpoint is universally accepted to be 30MHz. This may appear to be arbitrary, but it has in fact been found empirically that the coupling mechanisms are predominantly by conduction below 30MHz, and by radiation above it.

The regulation of emissions is intended only to reduce the threat to innocent "victim" receivers. Some reasonable separation of the offending emitter and the victim is assumed. If you place a personal computer right next to a domestic radio set you can still expect there to be interference. Regulations have nothing to say on the subject of intra-system interference, so that it is quite possible for two computer systems both of which meet the appropriate standard to be incompatible when they are placed together in the same rack or cabinet.

8.2   EMC legislation and standards

Up until the formulation of the European Community EMC Directive, the only countries to incorporate electromagnetic emission legislation in product requirements (except for household products) have been Germany and the United States.

Germany

In Germany, equipment operating at frequencies above 10kHz may not be used without a licence. A "General Licence" is granted to equipment which meets class B limits of the applicable standard. Alternatively a type-specific licence can be obtained for equipment which meets the class A limits. The user must ensure that the equipment they intend to use is covered by a licence, and so all manufacturers of equipment for the German market need to submit samples for type testing to the German VDE (Verband Deutscher Elektrotechniker) or an equivalent approved organisation. The applicable standards are VDE0875 for broadband interference (as generated by household appliances, etc.) and VDE0871 for broad- and narrow-band interference (as generated by information technology equipment for example).

United States

In the US, emissions from computing devices are governed by FCC (Federal Communications Commission) Rules part 15 subpart j. A "computing device" is any electronic device that generates or uses timing signals or pulses exceeding 10kHz and uses digital techniques. There are some quite broad exemptions from the rules depending on application. Two classes are defined, depending on the intended market: class A for business, commercial or industrial use, and class B for residential use. Like the VDE, these classes are subject to different limits, class B being the stricter. Before being able to market his equipment in the US, a manufacturer must either obtain certification approval from the FCC if it is a personal computer, or must verify that the device complies with the applicable limits.

8.2.1   The EMC Directive

The relaxed EMC regime that has hitherto existed throughout Europe, with the exception of Germany, is about to change dramatically. In accord with the general objective of the single European market, the European Commission has put forward an EMC directive whose purpose is to remove any barriers to trade on technical grounds relating to EMC. Thus, EC member states may not impede, for reasons relating to EMC, the free circulation on their territory of apparatus which satisfies the directive's requirements.

Scope and coverage

The directive applies to *all equipment placed on the market or taken into service*, so that it includes systems as well as individual products. It operates as follows: it sets out the essential requirements; it requires a statement to the effect that the equipment complies with these requirements; and it provides alternative means of determining whether the essential requirements have been satisfied.

The essential requirements are that

"The apparatus shall be so constructed that

(a) equipment shall not generate electromagnetic disturbances exceeding a level allowing radio and telecommunications equipment and other apparatus to operate as intended;

(b) equipment shall have an adequate level of intrinsic immunity from electromagnetic disturbances."

Thus protection is extended not only to radio and telecomms but also to other equipment such as information technology and medical equipment – in fact any equipment which is susceptible to electromagnetic (EM) disturbances. The second requirement states that the equipment should not malfunction in whatever hostile EM environment it may reasonably be expected to operate.

The range of EMC phenomena covered by the directive includes radiated emissions as well as those conducted along mains, signal, control or other cabling. The immunity requirement covers EM fields, spikes/dips/outages/distortions on the mains supply, electrostatic discharges, and lightning surges. Equipment within the scope of the directive can be considered under the following generic classifications:

industrial, scientific & medical; electricity supply and distribution; traction; lamps and luminaires; household appliances; building services; motor vehicles; broadcast; entertainment; ITE and telecomms; maritime; aeronautical; and military.

Routes to compliance

Many manufacturers will not be able to assess whether their equipment is able to satisfy the two essential requirements, so the directive looks towards the development of European standards on EMC. Any equipment which complies with the relevant standards will be deemed to comply with the essential requirements. However, a manufacturer may choose to undertake his own technical assessment – indeed, may have to if there is no relevant EMC standard in existence. In this case, he is required to keep a technical file containing details of the test method used, the test results and a supporting statement by an independent competent body (in the UK, a NAMAS accredited test house). This file must be at the disposal of the national administration.

In practice, the "technical file" route raises serious concerns (to the legislators) about how the Directive is applied, and how to determine when its requirements have not been met. These have yet to be resolved. A further procedure for demonstrating compliance is transitional and for the moment will apply for one year after the Directive comes into force, though this period may be extended later. It allows apparatus to continue to be governed by the national arrangements in force, in the absence of European or approved national standards.

The devlopment of generic standards

Clearly, there is an urgent need for the development of Europe-wide standards applicable to all electronic equipment and covering both emission of and immunity from interference. As we shall see shortly there are a number of product-oriented standards in existence, but a more productive approach is to have generic standards for the majority of equipment with a few product standards for specialised cases only. Such generic standards would comprise one main section giving levels of disturbance or immunity, together with several subsidiary parts detailing measurement methods.

The European standards body is CENELEC (the European Organisation for Electrotechnical Standardisation) which has UK representation from the BSI (the British Standards Institution). Once CENELEC has produced a European EMC standard all the CENELEC countries will be required to implement identical national standards, which will then be deemed to be "relevant standards" for the purpose of demonstrating compliance with the Directive.

A committee began work in 1989 on establishing the overall Europe-wide EMC standards. This committee set itself the task of generating such standards within two years, which, given that the average time to produce a European standard is six years, is extremely optimistic. Nevertheless, good progress is being made on the generic standards for the domestic, commercial and light industrial environment, and other parts of the generic standards are in the early draft stage at the time of writing.

Emission standards

Product sector	British	EN (CISPR)	FCC (US)	VDE (Germany)
Industrial, scientific & medical	BS4809	EN55011 (11)	Part 18	0871
Household appliances	BS800 ～ EN55014 (14)			0875
Fluorescent lighting	BS5394 ～ EN55015 (15)			0875
Radio & TV	BS905	EN55013 (13)		0872
Information technology equipment	BS6527 ～ EN55022 (22)		Part 15	0871

Immunity standards

Product sector	British	IEC	CISPR	Notes
Industrial process measurement & control	BS6667 ～ Pub. 801			RFI, ESD & transient susceptibility
Legal metrology	NWM 0320			RFI, ESD, transient & mains variations
Radio & TV receivers	BS905		Pub. 20	RFI only

Some British and European standards are aligned, these are indicated thus ～～～

Table 8.3 Summary of EMC standards

8.2.2   Existing standards

In the interim, manufacturers are looking to the clutch of product-oriented standards which already exist and attempting to apply them to their own products where this is possible and reasonable. Since emission standards have evolved over a number of years there is a good deal of agreement not only in methods of measurement but also in the limits themselves. Some of the principal standards are summarized in Table 8.3, and a comparison of the emission limits is shown in Figure 8.4 for conducted and Figure 8.5 for radiated emission. Measuring equipment is defined in CISPR publication 16, which is aligned with BS727 and which specifies the bandwidth and detector characteristics of the measuring receiver (Table 8.4), the impedance of the artificial mains network and

Parameter	Frequency Range		
•	9 to 150kHz	0.15 to 30MHz	30 to 1000MHz
Bandwidth	200Hz	9kHz	120kHz
Charge Time	45ms	1ms	1ms
Discharge Time	500ms	160ms	550ms
Overload factor	24dB	30dB	43.5dB

The CISPR16 quasi-peak detector response allows for the subjective variability in annoyance of pulse-type interference. For continuous (narrowband) interference the response is essentially that of a peak detector. For pulse interference, the receiver is progressively desensitized as the pulse repetition frequency is reduced.

Table 8.4 The CISPR16 quasi-peak measuring receiver

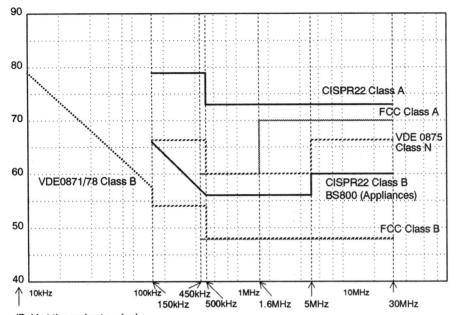

dBμV at the mains terminals
CISPR16 quasi-peak detector, 50Ω/50μH LISN

Figure 8.4 Conducted emission limits

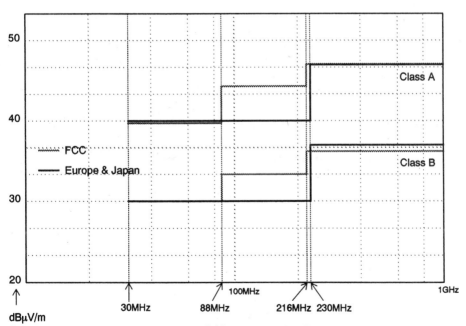

dBμV/m
CISPR16 quasi-peak detector, normalised (1/d) to a measuring distance of 10m

Figure 8.5 Radiated emission limits

NB consult current specifications for the up-to-date limit values

the construction of the coupling clamp.

Note that there are detailed differences in the methods of measurement allowed between the various emission standards. Measurements on conducted emissions below 30MHz are made on the mains terminals and use an artificial mains network, or Line Impedance Stabilising Network (LISN), to define the impedance of the mains supply. Radiated measurements above 30MHz require the use of an Open Area Test Site (OATS) to minimise reflections, with a calibrated attenuation characteristic. The limit levels shown in Figure 8.5 have been normalised to a measurement distance of 10m for ease of comparison; different standards call up distances of 3m, 10m or 30m. A direct translation of the levels (according to 1/d) from 10m to 3m is not strictly accurate because of the influence of near-field effects.

Immunity

Since immunity standards are of much more recent development they have not been adopted in the same way as emission standards. IEC Publication 801, aligned with BS6667, was published initially in 1984 and refers specifically to requirements for industrial process control instrumentation, but it includes methods of assessment which can be applied to a much wider range of products and it will almost certainly form the basis for the CENELEC generic immunity standards that will be required under the EMC Directive. At the time of writing it covers electrostatic discharge, RFI and fast transients, with a draft part in circulation covering switching and lightning surges. Another published immunity standard relates to legal metrology, where electronic methods are used for weighing and measuring products which are supplied to the general public. In this field the UK standard is the National Weights & Measures Laboratory (NWML) publication 0320. Also, CISPR publication 20 lays down requirements for the immunity of radio and TV receivers to RFI, primarily from citizens band transmitters, although it is being revised to cover a wider frequency range.

IEC 801 allows the choice of several test levels depending on the severity of the application. Because by contrast it is a legal standard, the CENELEC generic immunity standard will only have one test level. It will also have to find some way of defining acceptable performance criteria to evaluate the test results, but because of the diversity of equipment which it will cover this could prove elusive. These criteria will probably be based on loss of function or degradation in performance, but will have to distinguish between temporary, operator-recoverable and permanent failure.

8.3   Interference coupling mechanisms

As we have already discussed, interference can be coupled into or out of equipment over a number of routes (Figure 8.6). Chapter 1 (section 1.3) noted that electronic interactions follow the laws of electromagnetic field theory rather than the more convenient rules of circuit theory, and nowhere is this more evident than in EMC practice. At low frequencies the predominant modes of coupling are directly along the circuit wires or by magnetic induction, but at high frequencies each conductor, including the equipment housing if it is metallic, acts as an aerial in its own right and will contribute to coupling.

8.3.1   Conducted

A very frequent cause of conducted interference coupling is the existence of a common impedance path between the interfering and victim circuits. This path is usually though

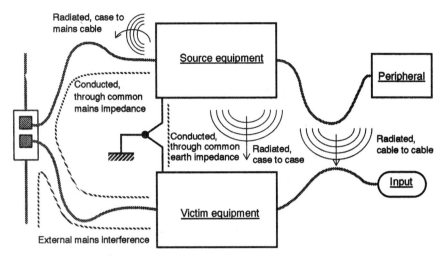

Figure 8.6 Coupling mechanisms

not invariably in the ground return. This topic has been covered extensively in chapter 1 and you are referred back to there for a full discussion. A typical case of common-impedance coupling might involve impulsive interference from a motor or switching circuit being fed into the 0V rail of a microprocessor circuit because they share the same earth return. Note that because of resonances, the impedance of earthing and bonding conductors is high at frequencies for which their length is an odd multiple of quarter-wavelengths.

Another coupling route is through the equipment power supply to or from the mains. The power supply forms the interface between the mains and the internal operation of the equipment, and as we have seen many emissions standards regulate the amount of interference that can be fed onto the mains via the power input leads. Power supply design was considered in more detail in chapter 7. A great deal of work has been done to characterise the impedance of the mains supply, and perhaps surprisingly measurements in quite different environments show close agreement. This has allowed

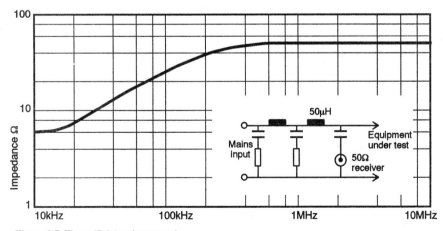

Figure 8.7 The artificial mains network

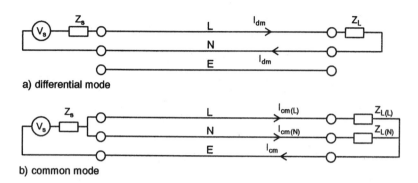

Figure 8.8 Interference propagation along power cables

the development of the CISPR16 artificial mains network (LISN) referred to in the last section. The impedance can be simulated by 50Ω in parallel with 50μH with respect to earth (Figure 8.7).

Power cables tend to act as low loss transmission lines up to 10MHz so that interference can propagate quite readily around the power distribution network, mainly attenuated by the random connection of other loads rather than by the cable itself. Interference can appear either as differential (symmetric) currents or as common mode (asymmetric) currents, as in Figure 8.8, and these need different treatment at the equipment, as we shall see when we discuss filters.

8.3.2   Radiated

When source and victim are near one another, radiated coupling is predominantly due either to magnetic or electric induction. Magnetic induction occurs when the magnetic flux produced by a changing current in the source circuit links with the victim circuit (*cf* section 1.1.4). The voltage induced in the victim circuit by a sinusoidal current I_s at frequency f due to a mutual inductance L henries is

$$V \quad = \quad 2\pi f \cdot I_s \cdot L \text{ volts}$$

Unfortunately calculating L accurately is difficult in most practical cases. It is proportional to the areas of the source and victim circuit loops, the distance between them, their relative orientation and the presence of any magnetic screening. As an example, short lengths of cable within the same wiring loom will have mutual inductances in the range of 0.1 to 3μH.

You will note that magnetic induction is a current (low-impedance) phenomenon, and it increases with increasing current in the source circuit. Electric induction, or capacitive coupling, is a voltage phenomenon and occurs when changing electric fields from the source interact with the victim circuit. The induced voltage due to a sinusoidal voltage V_s of frequency f on the source conductor coupled through a mutual capacitance C farads is

$$V \quad = \quad 2\pi f \cdot V_s \cdot C \cdot Z \text{ volts}$$

where Z is the impedance to ground of the victim circuit

Mutual capacitance between conductors depends on their distance apart, respective areas, and any dielectric material or electric screening between them. In some cases, component or cable manufacturers provide figures for mutual capacitance. It is generally in the range 1 – 100pF for typical circuit configurations.

Electromagnetic induction

When the source and victim are further apart then both electric and magnetic fields are involved in the coupling, and the conductors must be considered as antennas. For conductors whose dimensions are much smaller than the wavelength then the maximum electric field component in the far field at a distance d metres due to a current I at frequency f flowing in the conductor is

- for a loop,

$$E \quad = \quad 131.6 \cdot 10^{-16} \; (f^2 \cdot A \cdot I)/d \quad \text{volts per metre}$$

    where A is the area of the loop

- for a monopole against a ground plane,

$$E \quad = \quad 4\pi \cdot 10^{-7} \; (f \cdot L \cdot I)/d \quad \text{volts per metre}$$

    where L is the length of the conductor

As we saw at the beginning of this chapter, in the far field the field strength falls off linearly with distance. The electric and magnetic field strengths are related by the impedance of free space, $E/H = 377\Omega$. In the near field, a loop radiator will give a higher H (magnetic) field and this will fall off proportionally to $1/d^3$ while the E field falls off as $1/d^2$. Conversely, a short rod will give a high E field, which will fall off as $1/d^3$ while the H field falls off as $1/d^2$.

Once conductor lengths approach a quarter wavelength (one metre at 75MHz) then they cannot be treated as "electrically small" and they couple much more efficiently with ambient fields.

8.4   Circuit design and layout

A common response of circuit designers when they discover EMC problems late in the day is to specify some extra shielding and filtering in the hope that this will provide a cure. Usually it does, but this brute force approach may not be necessary if you put in some extra thought at the early circuit design stage. Shielding and filtering costs money, circuit design doesn't.

Most design for EMC is just good circuit design practice anyway. The most fundamental point to consider is the circuit's grounding regime: authorities on EMC agree that the majority of post-design interference problems can be traced to poor grounding. Printed circuit layout also has a significant impact. Chapter 1 considers ground design and chapter 2 relates this to pcbs and observations made there will not be repeated here, except to reiterate that short, direct tracks running close to their ground returns make very inefficient aerials and are therefore good for controlling both emissions and susceptibility.

8.4.1   Choice of logic

A careful choice of logic family will help to reduce high frequency emissions from digital equipment and may also improve rf and transient immunity. The harmonic spectrum of a trapezoidal wave, which approximates to a digital clock waveform,

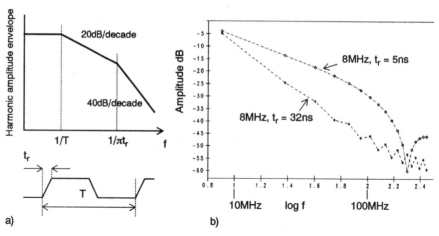

Figure 8.9 Harmonic amplitudes for a trapezoidal waveshape

shows a roll-off of amplitude with frequency which depends on the risetime (Figure 8.9 (a)). Using the slowest risetime compatible with reliable operation of your circuit will minimise the amplitude of the higher-order harmonics where radiation is more efficient. Figure 8.9(b) shows the calculated amplitude differences for an 8MHz clock with risetimes of 5ns and 32ns. An improvement approaching 20dB is possible at frequencies around 100MHz.

The advice based on this consideration is, use the slowest logic family that will do the job; don't use fast logic for no reason. Where parts of the circuit must operate at high speed, use fast logic only for those parts and keep the clocks local. If you are *in extremis*, try slugging the clock lines with a small parallel capacitor, but be very sure that all clock inputs can cope with the worst-case risetime that will result.

Noise margin and clock frequency

For good immunity, choose the logic family with the highest noise margin (see section 6.1.1). Spurious signals coupled into a logic signal circuit will have no effect until they reach the logic threshold. At the same time, the amplitude of signals coupled into the circuit from a given field will depend on the impedance of the circuit, which is defined by the driver output impedance. 4000B-series CMOS has around 10 times higher output impedance than LSTTL, depending on supply voltage (Table 6.1 on page 170), so although it has a higher input noise margin it is more susceptible to capacitively coupled interference. 74HC-series has a roughly equivalent output impedance and a higher input noise margin, so it is to be preferred over ordinary 74LS series TTL.

At the same time as you consider risetimes, also think about the actual clock frequency. Lowest is best. You may be able to use low frequency multi-phase clocks rather than a single high-frequency one. In some circumstances changing the spot frequency of the clock slightly may move its harmonics sufficiently far away from a particularly susceptible frequency, though this is more a case of EMC within a system than of meeting emission regulations.

8.4.2   Analogue circuits

As we have seen elsewhere (section 5.2.11), analogue circuits are also capable of unexpected oscillation at radio frequencies. Gain stages should be properly decoupled, loaded and laid out to avoid this. Check them at the prototype stage with a high-frequency 'scope or spectrum analyser, even if nothing appears to be wrong with the circuit function. Ringing on pulses transmitted along un-terminated transmission lines will generate frequencies which are related only to the length of the lines with perhaps sufficient amplitude to be troublesome. Terminate all long lines, particularly if they end at a CMOS input, which provides no inherent termination.

Good rf and transient immunity at interfaces calls for a consideration of signal bandwidth, balance and level. Any cable connecting to a piece of equipment will conduct interference straight into the circuit and it is at the interface that protection is needed. This is achieved at the circuit design level by a number of possible strategies:

- minimise the signal bandwidth, so that interfering signals outside the wanted frequency range are rejected;

- operate the interface at the highest possible power or voltage level consistent with other requirements, such as dynamic range, so that relatively more interference power is required to upset it;

- operate the interface where possible with the signal balanced, so that interference is injected in common-mode and is therefore attenuated by the common-mode rejection of the input circuit;

- in severe cases, galvanically isolate the input with an opto-isolator or transformer coupling, so that the only route for the interference is via the stray coupling capacitance of the isolation components.

Not so obvious is the overload performance of the circuit. If the interference drives the circuit into non-linearity then it will distort the wanted signal, but if the circuit remains linear in the presence of interference then it may be filtered out in a later stage without ill effect. Thus, any circuit which has a good dynamic range and a high overload margin will also be relatively immune to interference.

8.4.3   Software

If your circuit incorporates a microprocessor with resident embedded software then use all the available software tricks to overcome likely data corruption. This will be due primarily to transients, but also to RFI. These are discussed in chapter 6 but are reiterated briefly here:

- incorporate a watchdog timer. Any microprocessor without some form of watchdog is inviting disaster when it is exposed to disruptive transients.

- type-check and range-check all input data to determine its reliability. If it is outside range, reject it.

- sample input data several times and either average it, for analogue data, or act only on two or three successive identical logic states, for digital data – this is similar to digital switch de-bouncing.

- incorporate parity checking and data checksums in all data transmission.

- protect data blocks in volatile memory with error detecting and correcting codes. How extensively you use this protection depends on allowable time and memory overheads.

- wherever possible rely on level- rather than edge-triggered interrupts.

- do not assume that programmable interface chips (PIAs, ACIAs, etc) will maintain their initialized set-up state forever. Periodically re-initialize them.

8.5 Shielding

If despite the best circuit design practices, your circuit still radiates unacceptable amounts of noise or is too susceptible to incoming radiated interference, the next step is to shield it. This involves placing a conductive surface around the critical parts of the circuit so that the electromagnetic field which couples to it is attenuated by a combination of reflection and absorption. The shield can be an all-metal enclosure if protection down to low frequencies is needed, but if only high frequency (> 30MHz) shielding will be enough then a thin conductive coating deposited on plastic is adequate.

Shielding effectiveness

How well a shield attenuates an incident field is determined by its shielding effectiveness, which is the ratio of the field at a given point before and after the shield is in place. Shielding effectiveness of typical materials differs depending on whether the electric or magnetic component of the field is considered. Shielding effectiveness below 20dB is considered minimal, between 20 – 80dB is average, and 80 – 120dB is above average. Above 120dB is unachievable by cost-effective measures. The perfect electric shield consists of a seamless box with no apertures made from a zero-resistance material. This is known as a Faraday cage and does not exist. Michael Faraday described his attempt to build one as follows:

> "I had a chamber built, being a cube of 12 feet, and copper wire passed along and across it in various directions, and supplied in every direction with bands of tin foil, that the whole might be brought into good metallic communication. I went into the cube and lived in it, and using lighted candles, electrometers and all other tests of electrical states, I could not find the least influence on them, though all the time the outside of the cube was powerfully charged, and large sparks and brushes were darting off from every part of its outer surface."[†]

Any practical shield will depart from the ideal of infinite attenuation because of two factors:

- it is not made of perfectly conducting material
- it includes apertures and discontinuities

Shielding effectiveness of a solid conductive barrier can be expressed as the sum of reflection, absorption, and re-reflection losses:

$$SE_{(dB)} \;=\; R_{(dB)} + A_{(dB)} + B_{(dB)}$$

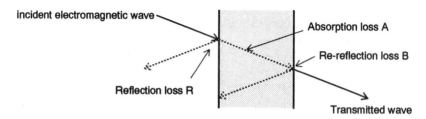

The reflection loss depends on the ratio of wave impedance to barrier impedance, the barrier impedance being a function of its conductivity and permeability, and of frequency. Reflection losses decrease with increasing frequency for the E-field

---

† **Experimental Researches in Electricity**, Michael Faraday, 1838, para 1173-1174

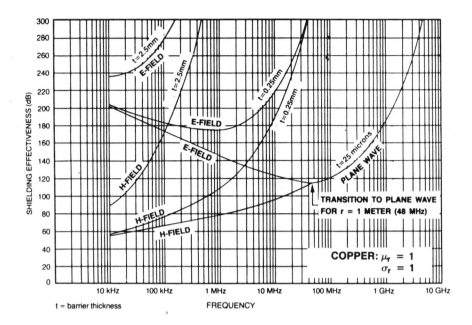

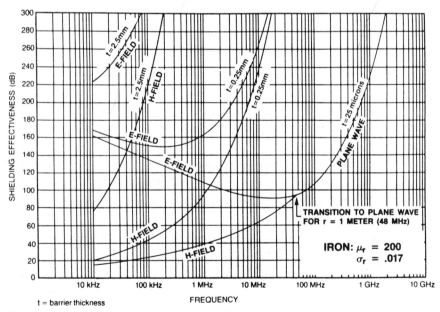

Figure 8.10 Composite reflection and absorption losses for copper and iron
Source: Tecknit

(electric) and increase for the H-field (magnetic). In the near field, closer than $\lambda/2\pi$, the distance between source and barrier also affects the reflection loss.

The re-reflection loss B is insignificant in most cases where A is greater than 10dB. A itself depends on the barrier thickness and its absorption coefficient. The inverse of the absorption coefficient is called the "skin depth" (δ). Skin depth is the measure of a magnetic phenomenon that tends to confine ac current to the surface of a conductor.

The skin depth reduces as frequency, permeability and conductivity increase, and fields are attenuated by 8.7dB (1/e) for every skin depth of penetration.

$$\delta \quad = \quad 6.61 \cdot (\mu_r \cdot \sigma_r \cdot f)^{-0.5} \text{ centimetres}$$
where μ_r is relative permeability (1 for air and copper), and σ_r is relative conductivity (1 for copper)

Thus at high frequencies, absorption loss becomes the dominant term. Figure 8.10 shows the combined reflection and absorption losses for copper and iron versus frequency.

8.5.1   Apertures

The curves in Figure 8.10 suggest that upwards of 200dB attenuation is easily achievable using reasonable thicknesses of common materials. In fact, the practical shielding effectiveness is limited by necessary apertures and discontinuities in the shielding.

Apertures are needed for ventilation, for control access, and for viewing indicators. Electromagnetic leakage through an aperture in a thin barrier depends on its longest dimension (d) and the minimum wavelength (λ) of the frequency band to be shielded against. For wavelengths less than or equal to twice the longest aperture dimension there is effectively no shielding. The frequency at which this occurs is the "cut-off frequency" of the aperture. For lower frequencies ($\lambda > 2d$) the shielding effectiveness increases linearly at a rate of 20dB per decade (Figure 8.11) up to the maximum possible for the barrier material. Comparing Figure 8.10 and Figure 8.11, you will see that for all practical purposes shielding effectiveness is determined by the apertures. For frequencies up to 1GHz (the present upper limit for radiated emissions standards) and a minimum shielding of 20dB the maximum hole size you can allow is 1.6cm.

For viewing windows in particular this is unacceptable and you then have to cover

Figure 8.11 Attenuation through an aperture

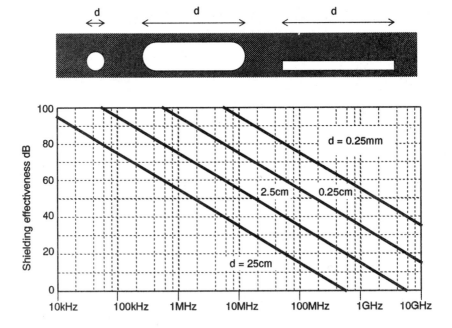

the window with a transparent conductive mesh, which must make good contact to the surrounding screen, or accept the penalty of shielding at lower frequencies only. You can cover ventilation holes with a perforated mesh screen without much trouble. If individual perforations are spaced closed together (hole spacing < $\lambda/2$) then the reduction in shielding over a single hole is approximately proportional to the square root of the number of holes. Thus a mesh of 100 4mm holes would have a shielding effectiveness 20dB worse than a single 4mm hole. Two similar apertures spaced greater than a half-wavelength apart do not suffer any significant extra shielding reduction.

8.5.2   Seams

An electromagnetic shield is normally made from several panels joined together at seams. Unfortunately, when two sheets are joined the electrical conductivity across the joint is imperfect. This may be because of distortion, so that surfaces do not mate perfectly, or because of painting, anodising or corrosion, so that an insulating layer is present on one or both metal surfaces (cf section 1.1.3). Consequently, the shielding effectiveness is reduced by seams just as much as it is by apertures. The reduction of shielding effectiveness depending on the longest dimension of the aperture applies equally to a non-conductive length of seam. The problem is especially serious for hinged front panels, doors and removable hatches that form part of a screened enclosure (Figure 8.12(a)). The penalty of poor contact is mitigated to some extent if the conductive sheets overlap, since this forms a capacitor which provides a partial current path at high frequencies. There are a number of other design options you can take to

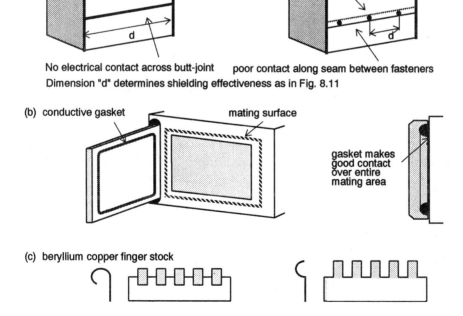

Figure 8.12 Improving conductivity of seams and joints

improve the shielding effectiveness at the seams:

- ensure that supposedly conductive surfaces remain conductive: don't paint or anodise them; alochrome is a suitable conductive finish for aluminium.

- maximise the area of overlap of two joined sheets. This can be done by lapped or flanged joints.

- where you are using screwed or riveted fastenings, space them as closely as possible. A good rule is no farther apart than $\lambda/20$, where λ is the wavelength at the highest frequency of interest.

- keep machining tolerances as tight as possible within your cost constraints. Where this still does not give a flat enough mating surface (which is likely to be most cases) or you need an environmental seal, or where you need good seam conductivity without fasteners such as at a hinged panel, use conductive gaskets (Figure 8.12(b)). These are available either as knitted wire mesh over rubber, or as an elastomer loaded with a conductive material such as silver flakes. Selection of the right gasket depends on environmental factors and the required conductivity.

- another way of improving shielding contact between two surfaces that frequently mate and un-mate is to use beryllium copper "finger" stock (Figure 8.12(c)) either as a continuous strip or spaced at intervals subject to the limitations on spacing given above for fasteners.

Above all, do not expect a supposedly shielded enclosure made of several ill-fitting painted panels with large holes and few fastenings to give you any serious shielding at all. To be fair, although as Figure 8.11 shows shielding effectiveness of such enclosures will be zero at uhf and above, in fact some equipment does not radiate sufficiently at these frequencies for shielding to be necessary to suppress radiated emissions. On the other hand, shielding against incoming uhf radio frequency interference will also be minimal.

8.6 Filtering

The purpose of filtering for EMC is almost invariably to attenuate high-frequency components while passing low-frequency ones. It is also, almost invariably, to block interfering signals which are coupled onto cables which enter or leave the equipment enclosure. There is little point in applying good shielding and circuit design practice to guard against radiated coupling if you then allow interference to be coupled in or out via the external connections. Interference can be induced directly onto the cable or can be coupled through the external connection, and a proper filter can guard against either or both of these, but you need some knowledge of the characteristics of the circuit in which the filter will be embedded to design or select the best filtering device.

Filters fall broadly into three categories: those intended for mains terminals, for input/output connections and for individual power or signal wires. The basic principle is the same for all of these.

8.6.1 The low-pass filter

This is not the place to go into the details of filter design, which are well covered in many books on circuit theory. The discussion here will confine itself to the practical aspects.

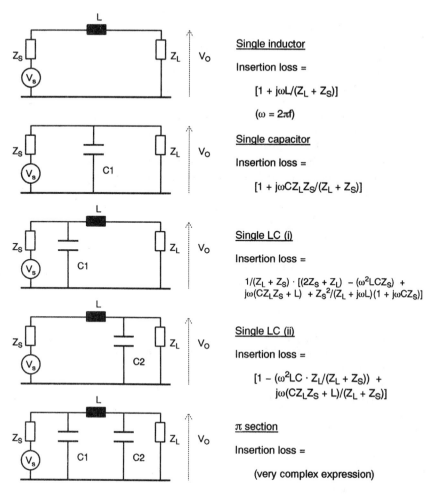

Figure 8.13 Low-pass filters

The simpler circuit arrangements which offer a low-pass response are shown in Figure 8.13. The attenuation of any given filter is conventionally quoted in terms of its insertion loss, that is the difference between the voltage across the load with the filter in and out of circuit. The first point to notice from the expressions to the right of the circuits in Figure 8.13 is that the insertion loss depends not only on the filter components but also on the source and load impedances. The second point to notice is that, except for the simplest circuits, it's a lot easier either to work out the theoretical insertion loss by computer on a circuit simulator, or to build the circuit and measure it!

Impedances

Knowledge of source and load impedances is fundamental. The simple inductor circuit, in which the inductor may be nothing more than a ferrite bead (see Figure 3.20) will give good results – better than 40dB attenuation – in a low impedance circuit but will be quite useless at high impedances. Conversely, the simple capacitor will give good results at high impedances but will be useless at low ones. The multi-component filters will give better results provided that they are configured correctly; the capacitor should face a high impedance and the inductor a low one.

Frequently, Z_S and Z_L are complex and perhaps unknown at the frequencies of interest for suppression. We have seen earlier (Figure 8.7) that the impedance of the mains supply is fairly predictable and can be quite easily modelled. You should be able to derive hf impedances for most signal circuits. It is not so easy for power supply inputs, as power components such as transformers, diodes and reservoir capacitors are not characterised at hf, and you will either have to measure the impedances or make an intelligent guess at the likely values. When the source is taken to be a cable acting as an antenna any analytical method for deriving its impedance is likely to be wildly inaccurate when applied to the real situation, in which cable orientation and positioning is uncontrolled, and a nominal value has to be assumed. 50Ω is usually specified as the test impedance for filter units, and this is as good a value as any to take for the external source impedance. You have to remember, though, that published insertion loss figures in a 50Ω system are not likely to be obtained in the real application. This doesn't necessarily matter, provided the filter-and-circuit combination is carefully characterised and tested as a whole.

Components and layout

Filter components, like all others, are imperfect. Inductors have self-capacitance, capacitors have self-inductance. This complicates the equivalent circuit at high frequencies and means that a typical filter using discrete components will start to lose its performance above about 10MHz. The larger the components are physically, the lower will be the break frequency. Chapter 3, section 3.3.9 discusses the self-resonance effects of capacitors, and transposing the impedance curves of Figure 3.17 to the low-pass filter circuits will show that as the frequency increases beyond capacitor self-resonance the impedance of the capacitors in the circuit actually rises, so that the insertion loss begins to fall. This can be countered by using special construction for the capacitors, as we shall see shortly when we look at feedthrough components.

The components are not the only cause of poor hf performance. Layout is another factor; lead inductance and stray capacitance can contribute as much degradation as component parasitics. Two common faults in filter applications are not to provide a decent low-inductance ground connection, and to wire the input and output leads in the same loom or at least close to each other, as in Figure 8.14.

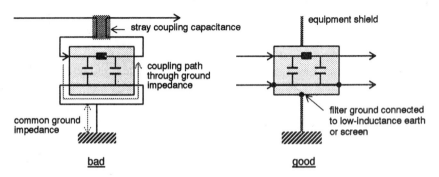

Figure 8.14 Filter wiring and layout

A poor ground offers a common impedance which rises with frequency and couples hf interference straight through from one side to the other via the filter's local ground path. Common input-output wiring does the same thing through stray capacitance. The cures are obvious: always mount the filter so that its ground node is directly coupled to the lowest inductance ground of the equipment, preferably the chassis. Keep the I/O leads separate, preferably screened from each other. The best solution is to position the filter so that it straddles the equipment shielding.

8.6.2 Mains filters

RFI filters for mains supply inputs have developed as a separate species and are available in many physical and electrical forms from several specialist manufacturers. A typical "block" filter for European mains supplies with average insertion loss might cost around £5. The reasons for this separate development are

- mandatory RFI emission standards have concentrated on interference conducted out of equipment via the mains, and consequently the market for filters to block this interference is large and well established, with predictable performance needs;

- there is an unfortunate tendency to add filtering as an afterthought, when it is discovered that equipment doesn't meet the regulations: add-on block mains filters are well matched to this "design" requirement;

- any components on the mains wiring side of equipment are exposed to an extra layer of safety regulatory requirements. Filter manufacturers are able to amortize the cost of designing and certifying their products to the plethora of national standards over a large number of units, thus relieving the equipment manufacturer to some extent of this particular burden;

- locating a filter directly at the mains inlet lends itself well to the provision of the whole input circuitry – connector, filter, fuse, on/off switch – as one block which the manufacturer can "fit and forget".

- many equipment designers are at a loss when it comes to rf filter design, and prefer a bought-in solution.

On the other hand, mains filters *can* be designed-in with the rest of the circuit, and this becomes a cost effective approach for high volume products and is almost always necessary if you need the optimum filter performance. A typical mains filter circuit (Figure 8.15) includes components to block both common-mode and differential-mode interference currents (*cf* Figure 8.8).

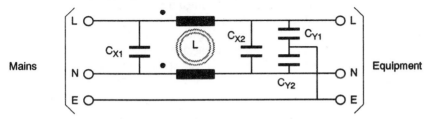

Figure 8.15 Typical mains filter circuit

The common-mode choke L consists of two identical windings on a single high permeability, usually toroidal, core, configured in the circuit so that differential (line-to-neutral) currents cancel each other. This allows high inductance values, typically 1–10mH, in a small volume without fear of choke saturation caused by the mains frequency supply current. The full inductance of each winding is available to attenuate common-mode currents with respect to earth, but only the leakage inductance, which depends critically on the choke construction, will offer attenuation to differential-mode interference.

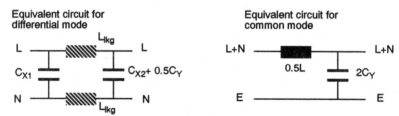

Figure 8.16 Filter equivalent circuits

Capacitors C_{X1} and C_{X2} attenuate differential-mode only but can have fairly high values, 0.1 to 0.47µF being typical. Either may be omitted depending on the detailed performance required, remembering that the source and load impedances may be too low for the capacitor to be useful. Capacitors C_{Y1} and C_{Y2} attenuate common-mode interference and if C_{X2} is large, have no significant effect on differential mode.

Safety requirements

The values of $C_{Y1,2}$ are nearly always limited by the allowable earth return current, which is set by safety considerations. This current is due to the operating voltage at mains frequency developed across the capacitors. Several national safety authorities define maximum earth current levels and these depend on the safety class of the equipment (see section 9.1.1) and on the actual application. Values range from 0.25mA to 5mA. BS613, which specifies requirements for mains rfi filters in the UK, gives the maximum value for Y-configured capacitors for class 1 appliances connected by a plug and socket as 0.005µF, and this value is frequently found in general purpose filters. The quality of the components is also critical, since they are continuously exposed to mains voltage; failure of either C_X or C_Y could result in a fire hazard, and failure of C_Y could also result in an electric shock hazard. You must therefore only use components which are rated for mains use in these positions (*cf* section 3.3.1).

Insertion loss versus impedance and current

Mains filter insertion loss is universally specified between 50Ω terminations. As we noted earlier, the actual in-circuit performance will be different because your circuit is unlikely to look like a flat 50Ω across the frequency range. It may also differ because of the working current. Increasing current through the inductor will eventually result in saturation, even of a dual wound common-mode choke; once the core saturates, the inductance falls dramatically and attenuation is lost. All commercially available filters have an rms current rating and provided they are operated within this rating there should be no problem. Unfortunately, typical power supply input currents show a high crest factor (see the chapter on power supplies) and the peak current may be 3 or more times the rms current. Although the filter may appear to be adequately rated on an rms basis, the peak current will overload it and it will be effectively useless.

8.6.3   I/O filters

In contrast to the mains filter, a filter on an I/O line has to be more closely tailored to individual applications, and consequently ready-made filters are not normally available. A major variable is the signal bandwidth of the I/O line. If the signal bandwidth extends into the rf range, as for example 10Mbit/s digital interfaces or video lines, then a simple low-pass filter cannot be used except for UHF and above. Conversely, a slow signal such as from a transducer or switch can easily be filtered with a simple capacitor.

A low-pass filter may affect the signal waveshape even if its cut-off frequency is higher than the signal bandwidth. More complex filter components with very steep cut-off characteristics are becoming available to address this problem.

I/O filters may also be required to clamp transients to a safe level, determined by the over voltage capability of the circuitry inboard of the filter. This is invariably achieved by a combination of low-pass filtering and transient suppression components, such as zener diodes (section 4.1.7) or varistors. A discrete component approach may suffice for many applications, but where fast rising transients are expected the lead and wiring inductance can have a significant effect on the circuit's ability to clamp the edge of the pulse. In these cases a combined capacitor/varistor component, in which the ceramic dielectric is treated to give it a predictable low-voltage breakdown characteristic, can offer a solution.

8.6.4   Feedthrough and 3-terminal capacitors

Any low-pass filter configuration except for the simple inductor uses a capacitor in parallel with the signal path. A perfect capacitor would give an attenuation increasing at a constant 20dB per decade as the frequency increased, but a practical wire-ended capacitor has some inherent lead inductance which in the conventional configuration puts a limit to its high frequency performance as a filter. The impedance characteristics given in Figure 3.17 show a minimum at some frequency and rise with frequency above this minimum. This lead inductance can be put to some use if the capacitor is given a three-terminal construction (Figure 8.17).

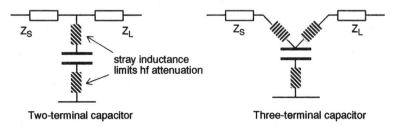

Two-terminal capacitor   Three-terminal capacitor

Figure 8.17 Two versus three terminals

The lead inductance now forms a T-filter with the capacitor, greatly improving its high-frequency performance. Lead inductance can be enhanced by incorporating a ferrite bead on each of the upper leads. The 3-terminal configuration can extend the effectiveness of a small ceramic capacitor from below 50MHz to upwards of 200MHz, which is particularly useful for interference in the vhf band.

Feedthroughs

Any leaded capacitor is still limited in effectiveness by the inductance of the connection to the ground point. For the ultimate performance, and especially where penetration of a screened enclosure must be protected at uhf and above (this is more often the case for military equipment) then a feedthrough construction is essential.

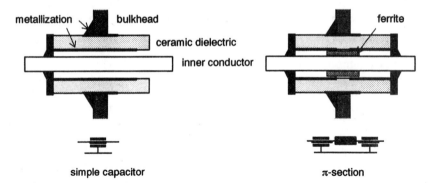

Figure 8.18 The feedthrough capacitor

Here, the ground connection is made by the outer body of the capacitor being screwed or soldered directly to the metal screening or bulkhead (Figure 8.18). There is effectively no inductance associated with this terminal and the capacitor performance is maintained well into the GHz region. The inductance of the through lead can be increased, thereby creating a π-section filter, by separating the ceramic metallization into two parts and incorporating a ferrite bead within the construction.

Feedthrough capacitors are available in a wide range of voltage and capacitance ratings but their cost increases with size. Cheap solder-in types between 100pF and 1000pF may be had for a few tens of pence, but larger screw mounting components will cost £1–£2. If you need good performance down to the low-MHz region then you will have to pay even more for it. A cheaper solution is to parallel a small feedthrough component with a larger, cheaper conventional unit which suppresses the lower frequencies at which physical construction is less critical.

Circuit considerations

When using any form of capacitive filtering, you have to be sure that your circuit can handle the extra capacitance to ground. This factor can be particularly troublesome when you need to filter an isolated circuit at radio frequencies. The rf filter capacitance provides a ready-made ac path to ground for the signal circuit and will seriously degrade the ac isolation, to such an extent that an rf filter may actually *increase* susceptibility to lower frequency common-mode interference. This is a function of the capacitance imbalance between the isolated signal and return lines (Figure 8.19), and it may restrict your allowable rf filter capacitance to a few tens of pF.

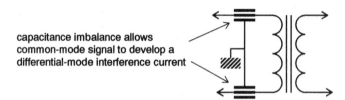

Figure 8.19 Unbalance between feedthrough capacitors

Another problem may arise if you are filtering several signal lines together and using a common earth point, as is the case for example with the filtered-D range of connectors. Provided that the earth connection is low impedance there is no problem, but any series impedance in the earth path not only degrades the filtering but will also couple signals from one line into another (Figure 8.20), leading to designed-in crosstalk. For this reason filtered connectors should be used with care, and they must always be well-grounded to the case – and make sure that the case is also the signal ground!

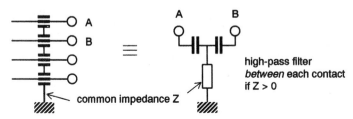

Figure 8.20 Common impedance coupling through filter capacitors

8.7   Cables and connectors

Cabling has already been discussed in chapter 1, but we will look at it again briefly here, since the EMC of any given product is always affected by the configuration of the cables that are connected to it.

In the presence of an electromagnetic field any cable acts as an antenna and energy from the field is coupled onto it. The current induced on the cable depends on its physical orientation with respect to the field and any nearby conductive objects, and on its length. In a reciprocal manner, the field radiated by a cable carrying an rf current also depends on these parameters. Cable length is sometimes under your control, but orientation never is (unless you are a system designer). Therefore, you need to take steps to prevent interfering cable currents from affecting circuit operation, or to prevent circuit operation from generating interfering cable currents.

Properly terminating the cable shield

One approach is to filter the signal lines at the point at which they enter or leave the cable, as we saw in the previous section. Where this is inadequate or impossible, the other approach is to surround the signal conductors with a conductive shield which is grounded to the equipment screen, as is shown in Figure 1.17 on page 18. The function of this shield is to provide a return path for induced currents which does not couple onto the signal conductors, or conversely to confine radiating currents that are present on the signal conductors and prevent them from coupling with external fields. Figure 1.17 shows the ideal connection of a shield to screen in which there is no discontinuity between the two. This can be most nearly achieved when the cable is fixed to the equipment and led through a conductive gland so that the cable screen makes contact all the way round its circumference, through 360°. As soon as a connector is used, some compromise has to be made.

Military-style connectors are designed as above so that the cable screen makes 360° contact, but they attract military-style prices and assembly costs. RF coaxial connectors, such as the common BNC type, also make 360° contact , but they carry only one signal line at a time. Most multi-way connectors do not have proper provision for terminating the shield, and this is where performance degradation creeps in. Only too often, the shield is brought down to a "pig-tail" or drain wire and terminated to one of the connector pins – the EIA/RS-232D interface standard (section 6.2.5) even has a pin allocated to this function, pin 1 – or, worse still, it is not terminated at all (Figure 8.21).

The effect of no shield connection at all is to nullify the shielding effectiveness at high frequency. This is perhaps to be expected, but what is less intuitively obvious is that the pigtail connection is almost as bad. The difference in effectiveness between a pigtail connection and a full 360° connection is minimal below 3MHz, but can approach 40dB at higher frequencies. It is caused by resonant effects on the pigtail inductance, and can show variations greater than 20dB over small changes in frequency.

Screened backshells

The best termination for a multi-way connector is to use a screened conductive backshell for the connector and to clamp the cable shield firmly to it. The backshell must make contact directly with the conductive shell of the connector itself, and this in turn must make good 360° contact with the shell on its mating connector, which must be bolted firmly and conductively to the equipment case. Any departure from this practice – in terms of not bolting directly to the case, not having mating conductive shells on the connectors, and not using a screened backshell – will compromise the high-frequency shielding of the system.

Inexpensive subminiature "D"-type multi-way connectors are now available whose construction adheres to these principles, and if they are used intelligently they can give good screening. It is quite possible to misuse them either in design or assembly and throw away all their advantages. Most other types of multi-way connector, in particular the popular insulation displacement two-part units, have no potential for making

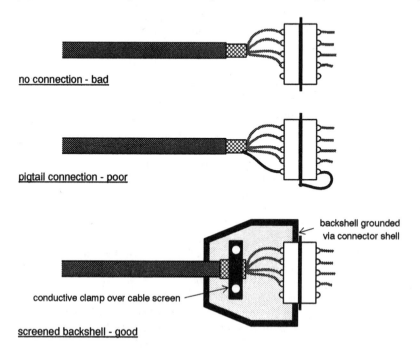

no connection - bad

pigtail connection - poor

backshell grounded
via connector shell

conductive clamp over cable screen

screened backshell - good

Figure 8.21 Cable screen terminations

shielded connections at all and should only be used for inter-board connections inside screened equipment. The challenge of a cheap, easy-to-assemble and effective screened cable/connector system has still to be met.

8.8   EMC design checklist

- Design for EMC from the beginning; know what performance you require
- Select components and circuits with EMC in mind:
    - use slow and/or high-immunity logic
    - use good rf decoupling
    - minimise signal bandwidths, maximise levels
    - provide power supplies of adequate (noise-free) quality
    - incorporate a watchdog circuit on every microprocessor
- PCB layout:
    - ensure proper signal returns; if necessary include isolation to define preferred current paths
    - keep interference paths segregated from sensitive circuits
    - minimise ground inductance with thick gridding or ground plane
    - minimise loop areas in high-current or sensitive circuits
    - minimise track and component leadout lengths
- Cables:
    - avoid parallel runs of signal and power cables
    - use signal cables and connectors with adequate screening
    - use twisted pair if appropriate
    - run cables away from apertures in the shielding
    - avoid resonant lengths where possible
- Grounding:
    - ensure adequate bonding of screens, connectors, filters, cabinets etc
    - ensure that bonding methods will not deteriorate in adverse environments
    - mask paint from any intended conductive areas
    - keep earth leads short
    - avoid common ground impedances
- Filters:
    - optimise the mains filter for the application
    - use correct components and filter configuration for I/O lines
    - ensure a good ground return for each filter
    - apply filtering to interference sources, such as switches or motors
- Shielding:
    - determine the type and extent of shielding required from the frequency range of interest
    - enclose particularly sensitive or noisy areas with extra internal shielding
    - avoid large or resonant apertures in the shield, or take measures to mitigate them
- Test and evaluate for EMC continuously as the design progresses

Chapter 9

General product design

9.1 Safety

Any electronic equipment must be designed for safe operation. Most countries have some form of product liability legislation which puts the onus on the manufacturer to ensure that his product is safe. The responsibility devolves onto the product design engineer, to take reasonable care over the safety of the design. This includes ensuring that the equipment is safe when used properly, that adequate information is provided to enable its safe use, and that adequate research has been carried out to discover, eliminate or minimize risks due to the equipment.

There are various standards relating to safety requirements for different product sectors. In some cases, compliance with these standards is mandatory. In the European Community, the Low Voltage Directive (73/23/EEC) applies to all electrical equipment with a voltage rating between 50 and 1000Vac or 75 and 1500Vdc, with a few exceptions, and requires member states to take all appropriate measures

> "to ensure that electrical equipment may be placed on the market only if, having been constructed in accordance with good engineering practice in safety matters in force in the Community, it does not endanger the safety of persons, domestic animals or property when properly installed and maintained and used in applications for which it was made."

If the equipment conforms to a harmonized CENELEC or internationally-agreed standard then it is deemed to comply with the Directive. An example of a harmonized standard is BS415 : 1990, "Safety requirements for mains-operated electronic and related apparatus for household and similar general use", which is equivalent except for certain special national conditions with IEC Publication 65 of the same title. Proof of compliance can be by a Mark or Certificate of Compliance from a recognized laboratory, or by the manufacturer's own declaration of conformity. The Directive includes no requirement for compulsory approval for electrical safety.

The hazards of electricity

The chief dangers (but by no means the only ones, see Table 9.1) of electrical equipment are the risk of electric shock, and the risk of a fire hazard. The threat to life from electric shock depends on the current which can flow in the body. For ac, currents less than 0.5mA are harmless, whilst those greater than 50-500mA (depending on duration) can be fatal.[†] Protection against shock can be achieved simply by limiting the current to a safe level, irrespective of the voltage. There is an old saying, "it's the volts that jolts, but the mils that kills". If the current is not limited, then the voltage level in conjunction with contact and body resistance determines the hazard. A voltage of less

---

† IEC Publication 479 : 1984 gives further information.

Hazard	Main risk	Source
Electric shock	Electrocution, injury due to muscular contraction, burns	Accessible live parts
Heat or flammable gases	Fire, burns	Hot components, heatsinks, damaged or overloaded components and wiring
Toxic gases or fumes	Poisoning	Damaged or overloaded components and wiring
Moving parts, mechanical instability	Physical injury	Motors, parts with inadequate mechanical strength, heavy or sharp parts
Implosion/explosion	Physical injury due to flying glass or fragments	CRTs, vacuum tubes, overloaded capacitors and batteries
Ionizing radiation	Radiation exposure	High-voltage CRTs, radioactive sources
Non-ionizing radiation	RF burns, possible chronic effects	Power RF circuits, transmitters, antennas
Laser radiation	Damage to eyesight, burns	Lasers
Acoustic radiation	Hearing damage	Loudspeakers, ultrasonic transducers

Table 9.1 Some safety hazards associated with electronic equipment

than 50V ac rms, isolated from the supply mains or derived from an independent supply, is classified as a Safety Extra-Low Voltage (SELV) and equipment designed to operate from an SELV can have relaxed requirements against the user being able to contact live parts.

Aside from current and voltage limiting, other measures to protect against electric shock are

- earthing, and automatic supply disconnection in the event of a fault. See section 1.1.12.

- Inaccessibility of live parts. A live part is any part, contact with which may cause electric shock, that is any conductor which may be energized in normal use – not just the mains "live".

9.1.1   Safety classes

BS2754 (related to IEC publication 536) classifies electrical equipment into four classes according to the method of connection to the electrical supply and gives guidance on forms of construction to use for each class. The classes are

Class 0:   Protection relies on basic functional insulation only, and there is no provision for an earth connection. This construction is unacceptable in the UK.

Class I: Equipment is designed to be earthed. Protection is afforded by basic insulation, but failure of this insulation is guarded against by bonding all accessible conductive parts to the protective earth conductor. It depends for its safety on a satisfactory earth conductive path being maintained for the life of the equipment.

Class II: The equipment has no provision for protective earthing and protection is instead provided by additional insulation measures, such as double or reinforced insulation. Double insulation is functional insulation, plus a supplementary layer of insulation to provide protection if the functional insulation fails. Reinforced insulation is a single layer which provides equivalent protection to double.

Class III: Protection relies on supply at safety extra low voltage and voltages higher than SELV are not generated. Second-line defences such as earthing or double insulation are not required.

9.1.2   Design considerations for safety protection

The requirement for inaccessibility has a number of implications. Any openings in the equipment case must be small enough that the British Standard test finger, whose dimensions are defined in those standards that call up its use, cannot contact a live part (Figure 9.1). Worse, small suspended bodies (such as a necklace) that can be dropped through ventilation holes must not become live. This may force the use of internal baffles behind ventilation openings.

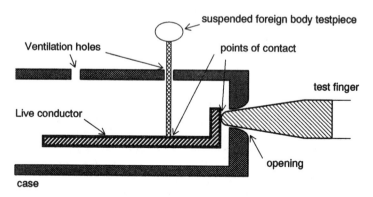

Figure 9.1 The test finger and the suspended foreign body

Protective covers, if they can be removed by hand, must not expose live parts. If they do, they must only be removable by use of a tool. Or, use extra internal covers over live portions of the circuit. It is anyway good practice to segregate high-voltage and mains sections from the rest of the circuit and provide them with separate covers. Most electronic equipment runs off voltages below 50V and, provided the insulation offered by the mains isolating transformer is adequate, the signal circuitry can be regarded as being at SELV and therefore not live.

Any insulation must, in addition to providing the required insulation resistance and dielectric strength, be mechanically adequate. It will be dropped, impacted, scratched and perhaps vibrated to prove this. It must also be adequate under humid conditions: hygroscopic materials (those that absorb water readily, such as wood or paper) are out.

Various standards define acceptable creepage and clearance distances versus the voltage proof required. As an example, BS415 allows 0.5mm below 34V rising to 3mm at 354V and extrapolated thereafter; distances between PCB conductors are slightly relaxed, being 0.5mm up to 124V, increasing to 3mm at 1240V. Creepage distance (Figure 9.2) denotes the shortest distance between two conducting parts along the surface of an insulating material, while clearance distance denotes the shortest distance through air.

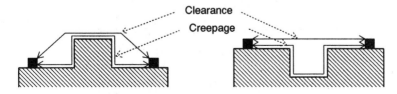

Figure 9.2 Creepage and clearance distance

Easily discernible, legible and indelible marking is required to identify the apparatus and its mains supply, and any protective earth or live terminals. Mains cables and terminations must be marked with a label to identify earth, neutral and live conductors, and class I apparatus must have a label which states "WARNING: THIS APPARATUS MUST BE EARTHED". Fuse holders should also be marked with their ratings and mains switches should have their "off" position clearly shown. If user instructions are necessary for the safe operation of the equipment, they should preferably be marked permanently on the equipment.

Any connectors which incorporate live conductors must be arranged so that exposed pins are on the dead side of the connection when the connector is separated. When a connector includes a protective earth circuit, this should mate before the live terminals and unmate after them. (See the CEE-22 6A connector for an example).

9.1.3   Fire hazard

It is taken for granted that the equipment won't overheat during normal operation. But you must also take steps to ensure that it does not overheat or release flammable gases to the extent of creating a fire hazard under fault conditions. Any heat developed in the equipment must not impair its safety. Fault conditions are normally taken to mean short-circuits across any component, set of terminals or insulation that could conceivably occur in practice (creepage and clearance distances are applied to define whether a short circuit would occur across insulation), stalled motors, failure of forced cooling and so on.

The normal response of the equipment to these types of faults is a rise in operating current, leading to local heating in conductors. The normal protection method is by means of fuses, thermal cutouts or circuit breakers in the supply or at any other point in the circuit where over-current could be hazardous. As well as this, flame-retardant materials should be used wherever a threat of overheating exists, such as for pcb base laminates.

Fuses are cheap and simple but need careful selection in cases where the prospective fault current is not that much higher than the operating current. They must be easily replaceable, but this makes them subject to abuse from unqualified users (hands up anyone who hasn't heard of people replacing fuselinks with bent nails or pieces of cigarette-packet foil). The manufacturer must protect his liability in these cases by clear labelling of fuseholders and instructions for fuse replacement. Fuse specification is covered in more detail in section 7.2.3.

Thermal cutouts and circuit breakers are more expensive, but offer the advantage of easy resetting once the fault has cleared. Thermal devices must obviously be mounted in close thermal contact with the component they are protecting, such as a motor or transformer.

9.2   Design for Production

Really, every chapter in this book has been about design for production. As was implied in the introduction, the ability which marks out a professional designer is the ability to design products or systems which work under all circumstances and which can be manufactured easily.

The sales and marketing engineer addresses the questions, "can I sell this product?" and "how much can I sell this product for?" This book hasn't touched on these issues, important though they are to designers; it has assumed that you have a good relationship with your marketing department and that your marketing colleagues are good at their job. But you as designer also have to address another set of questions, which are

- can the purchasing department source the components quickly and cheaply?
- can the production department make the product quickly and cheaply?
- can the test department test it easily?
- can the installation engineers or the customer install it successfully?

It is as well to bear all these questions in mind when you are designing a product, or even part of one. Your company's financial health, and consequently your and others' job security, ultimately depends on it. A good way to monitor these factors is to follow a checklist.

9.2.1   Checklist

Sourcing

- Have you involved purchasing staff as the design progressed?
- Are the parts available from several vendors or manufacturers wherever possible? Have you made extensive use of industry standard devices?
- Where you have specified alternate sources, have you made sure that they are all compatible with the design?
- Have you made use of components which are already in use on other products?
- Have you specified close-tolerance components only where absolutely necessary?
- Where sole-sourced parts have to be used, do you have assurances from the vendor on price and lead time? How reliable are they? Have you checked that there is no warning, "not recommended for new designs", on each part?
- Does your company have a policy of vetting vendors for quality control? If so, have you added new vendors with this product, and will they need to be vetted?

Production

- Have you involved production staff as the design progressed?
- Are you sure that the mechanical and electrical design will work with all mechanical and electrical

tolerances?

- Does the mechanical design allow the component parts to be fitted together easily?

- Are components, especially polarised ones, all oriented in the same direction on the pcb for ease of inspection and insertion?

- Are discrete components, notably resistors, capacitors and transistors, specified to use identical pitch spacings and footprints as far as possible?

- Have you minimised wiring looms to front or rear panels and between pcbs, and used mass-termination connections (e.g. IDC) wherever possible?

- Have you modularised the design as far as possible to make maximum use of multiple identical units?

- Is the soldering and assembly process (wave, infra-red, auto-insert etc) that you have specified compatible with the manufacturing capability?

- If the production calls for any special assembly procedures (e.g. potting or conformal coating), or if any components require special handling or assembly (MOSFETs, LEDs, batteries, relays etc) are the production and stores staff fully conversant with these procedures and able to implement them? Have you minimised the need for such special procedures?

- Do all pcbs have adequate solder mask, track and hole dimensions, clearances, and silk screen legend for the soldering and assembly process? Are you sure that the test and assembly personnel are conversant with the legend symbols?

- Are your assembly drawings clear and easy to follow?

Testing and calibration

- Have you involved test staff as the design progressed?

- Are all adjustment and test points clearly marked and easily accessible?

- Have you used easily-set parts such as DIL switches or linking connectors in preference to solder-in wire links?

- Does the circuit design allow for the selection of test signals, test subdivision and stimulus/response testing where necessary?

- If you are specifying automatic testing with ATE, does the pc layout allow adequate access and tooling holes for bed-of-nails probing? Have you confirmed the validity of the ATE program and the functional test fixture?

- Have you written and validated a test software suite for microprocessor-based products?

Installation

- Is the product safe?

- Does the design have adequate EMC?

- Are the installation instructions or user handbook clear, correct and easy to follow?

- Do the installation requirements match the conditions which will obtain on installation? E.g., is the environmental range adequate, the power supply appropriate, the housing sufficient, etc..?

9.2.2   The dangers of ESD

There is one particular danger to electronic components and assemblies that is present in both the design lab and the production environment. This is damage from electrostatic discharge (ESD). This can cause complete component failure, as was discussed in section 4.5.1, or worse, performance degradation that is difficult or impossible to detect. It can also cause transient malfunction in operating systems (*cf* page 235).

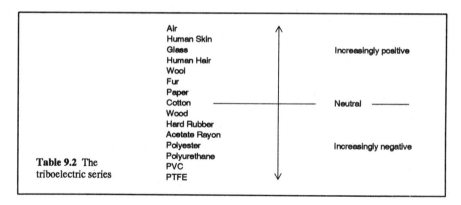

Table 9.2 The triboelectric series

Generation of ESD

When two non-conductive materials are rubbed together, electrons from one material are transferred to the other. This results in the accumulation of *triboelectric* charge on the surface of the material. The amount of the charge caused by movement of the materials is a function of the separation of the materials in the triboelectric series, shown in Table 9.2. Additional factors are the closeness of contact, rate of separation and humidity. Figure 8.3 on page 236 shows the electrostatic voltage related to materials and humidity, from which you can see that possible voltages can exceed 10kV.

If this charge is built up on the human body, as a result of natural movements, it can then be discharged through a terminal of an electronic component. This will damage the component at quite low thresholds, easily less than 1kV, depending on the device. Of the several contributory factors, low humidity is the most severe; if relative humidity is higher than 65% (which is frequent in maritime climates such as the UK's) then little damage is likely. Lower than 20%, as is common in continental climates such as the United States, is much more hazardous.

Gate-oxide breakdown of MOS or CMOS components is the most frequent, though not the only, failure mode. Static-damaged devices may show complete failure, intermittent failure or degradation of performance. They may fail after one very high voltage discharge, or because of the cumulative effect of several discharges of lower potential.

Static protection

To protect against ESD damage, you need to prevent static buildup and to dissipate or neutralize existing charges. At the same time, operators (including yourself and your design colleagues) need to be aware of the potential hazard. The methods used to do this are

- package sensitive devices or assemblies in conductive containers, keep them in these until use and ensure they are clearly marked

- use conductive mats on the floor and workbench where sensitive devices are assembled, bonded to ground via a 1MΩ resistor

- remove non-conductive items such as polystyrene cups, synthetic garments, wrapping film etc from the work area

- ground the assembly operator through a wrist strap, in series with a 1MΩ resistor for electric shock protection

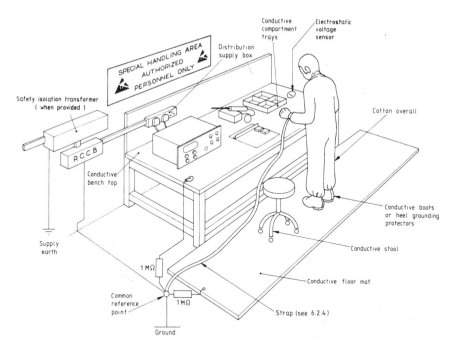

Figure 9.3 Static-safe workshop layout
Extract from BS5783:1987 reproduced with the permission of BSI

- ground soldering tool tips
- use ionized air to dissipate charge from non-conductors, or maintain a high relative humidity
- create and maintain a static-safe work area where these practices are adhered to
- ensure that all operators are familiar with the nature of the ESD problem
- mark areas of the circuit where a special ESD hazard exists; design circuits to minimise exposed high-impedance or unprotected nodes

All production assembly areas should be divided into static-safe workstation positions. Your design and development prototyping lab should also follow these precautions, since it is quite possible to waste considerable time tracking down a fault in a prototype which is due to a static-damaged device. A typical static-safe area layout is shown in Figure 9.3. BS5783 gives a code of practice for the handling of electrostatic sensitive devices.

9.3 Testability

The previous section's checklist included some items which referred to the testability of the design. It is vital that you give sufficient thought throughout the design of the product as to how the assembled unit or units will be tested to prove their correct function. In the very early stages, you should already know whether your test department will be using in-circuit testing, manual functional testing or functional testing on ATE, or a combination of these methods. You should then be in a position to include test access points and circuits in the design as it progresses. This is a far more cost-effective way of incorporating testability than merely bolting it on at the end.

9.3.1   In-circuit testing

The first test for an assembled pcb is to confirm that every component on it is correctly inserted, of the right type or value, and properly soldered in. It is quite possible for manual assembly personnel to insert the wrong component, or insert the right one incorrectly polarised, or even to omit a component or series of components. Automatic assembly is supposed to avoid such errors, but it is still possible to load the wrong component into the machine, or for components to be marked incorrectly. Automatic soldering has a higher success rate than hand soldering but bad joints due to lead or pad contamination can still occur.

In-circuit testing lends itself to automatic test fixture and test program generation. Each node on the pcb has to be probed, which requires a bed-of-nails test fixture (Figure 9.4) and this can be designed automatically from the pcb layout data. Similarly, the expected component characteristics between each node can be derived from the circuit schematic, using a component parameter library.

An in-circuit tester carries out an electrical test on each component, to verify its behaviour, value and orientation, by applying voltages to nodes that connect to each component and measuring the resulting current. Interactions with other components are prevented by guarding or back-driving. The technique is successful for discrete components but less so for integrated circuits, whose behaviour cannot be described in terms of simple electrical characteristics. It is therefore most widely applied on boards which contain predominantly discretes, and which are produced in high volume, as the overhead involved in programming and building the test fixture is significant. It does not of itself guarantee a working pcb. For this, you need a functional test.

9.3.2   Functional testing

A functional test checks the behaviour of the assembled board against its functional specification, with power applied and with simulated or special test signals connected to the input/output lines. It is often combined with calibration and set-up adjustments. For low-volume products you will normally write a test procedure around individual

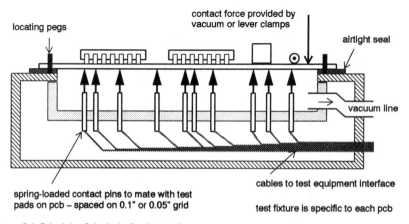

Figure 9.4 Principle of the bed-of-nails test fixture

test instruments, such as voltmeters, oscilloscopes and signal generators. You may go so far as to build a special test jig to simulate some signals, interface others, provide monitored power and make connections to the board under test. The test procedures will consist of a sequence of instructions to the test technician – apply voltage A, observe signal at B, adjust trimmer C for a minimum at D, and so on – along with limit values where measurements are made.

The disadvantage of this approach is that it is costly in terms of test time. This puts up the overhead cost of each board and affects the final cost price of the overall unit. It is cheap as far as instrumentation goes, since you only need a simple test jig, and you will normally expect the test department to have the appropriate lab equipment to hand. Hence it is best suited to low production volumes where you cannot amortize the cost of automatic test equipment.

A further, hidden, disadvantage may be that you don't have to define the testing absolutely rigorously but can rely on the experience of the test technician to make good any deficiencies in the test procedure or measurement limits. It is common for test personnel to develop a better "feel" for the quirks of a particular design's behaviour under test than its designer ever could. Procedural errors and invalid test limits may be glossed over by a human tester, and if such information is not fed back to the designer then the opportunity to optimise that or subsequent designs is lost.

ATE

Functional testing may alternatively be carried out by automatic test equipment (ATE). In this case, the function of the human tester is reduced to that of loading and unloading the unit under test, pressing the "go" button and observing the pass/fail indicator. The testing is comprehensively de-skilled; the total unit test time is reduced to a few minutes or less. This minimises the test cost.

The costs occur instead at the beginning of the production phase, in programming the ATE and building a test fixture. The latter is similar to (in some cases may be identical to) the bed-of-nails fixture which would be used for in-circuit testing (Figure 9.4). Or, if all required nodes are brought out to test connectors, the test fixture may consist of a jig which automatically connects a suite of test instrumentation to the board under the command of a computer-based test program. The IEEE-488 standard bus allows interconnection of a desktop computer and remote-controlled meters, signal generators and other instruments, for this purpose.

The skill required of a test technician now resides in the test program, which may have been written by you as designer or by a test engineer. In any case it needs careful validation before it is let loose on the product, since it does not have the skill or expertise to determine when it is making an invalid test. The cost involved in designing and building the test fixture, programming it and validating the program, and the capital cost of the ATE itself need to be carefully judged against the savings that will be made in test time per unit. It is normally only justified if high production volumes are expected.

9.3.3   Design techniques

There are many ways in which you can design a pcb circuit to make it easy to test, or conversely hard to test. The first step is to decide how the board will be tested, which is determined by its complexity, expected production volume and the capabilities of the test department.

Bed-of-nails probing

If you will be using a bed-of-nails fixture, then the pcb layout should allow this. Leave a large area around the outside of the board, and make sure there are no unfilled holes, to enable a good vacuum pressure to be developed to force the board onto the probes. Or if the board will be clamped to the fixture, make sure there is space on the top of the board for the clamps. Decide where your test nodes need to be electrically, and then lay out the board to include target pads on the underside for the probes. These pads should be spaced on a 0.1" grid for accurate drilling of the test jig; 0.05" is possible but fiddly and unreliable. It is not good practice to use component lead pads as targets, since pressure from the probe may cause a defective joint to appear good. Ensure that tooling holes are provided and are accurately aligned with the targets.

Remember that a bed-of-nails jig will connect several long, closely coupled wires to many nodes in the circuit. This will severely affect the circuit's stray reactances, and thereby modify its high frequency response. It is not really suitable for functionally testing high frequency or high speed digital circuits.

Test connections

If your test department doesn't want to use bed-of-nails probing, then help them find the test points that are necessary by bringing them out to test connectors. These can be cheap-and-cheerful pin strips on the board since they will normally only be used once. The matching test jig can then take signals from these connections direct to the test instrumentation via a switch arrangement. Of course, pre-existing connectors such as multi-way or edge connectors can be used to bring out test signals on unused pins. Be careful, though, that you do not bring long test tracks from one side of the board to the other and thereby compromise the circuit's crosstalk, noise susceptibility and stability. Extra local test connectors are preferable.

Circuit design

There are many design tricks to make testing easier. A simple one is to include a series resistor in circuits where you will want to back-drive against an output, or where you will want to measure a current (Figure 9.5(a)). The cost of the resistor is minimal compared to the test time it might save. Of course, you must ensure that the resistor does not affect normal circuit operation. Also, unused digital gate inputs may be taken to a pull-up resistor rather than direct to supply or ground (cf section 6.1.5), and this point can then be used to inhibit or enable logic signals for testing purposes only (Figure 9.5(b)).

The theme of Figure 9.5(b) can be taken further to incorporate extra logic switching to allow data or timing signals to be derived either from the normal on-board source, or from the external test equipment. This is particularly useful in situations where testing logic functions from the normal system clock would result either in too fast operation, or too slow. The clock source can be taken through a 2-input data multiplexer such as the 74HC157, one input of which is taken via a test connector to the external clock as shown in Figure 9.6. In normal operation the clock select and test clock inputs are left unconnected and the system clock is passed directly through the multiplexer.

When you are considering testing a microprocessor board, it is advantageous to have a small suite of test software resident on the main program PROM. This can be activated on start-up by reading a digital input which is connected to a test link or test probe. If the test input is set, the program jumps to the test routines rather than to the main operating routines. These are arranged to exercise all inputs, outputs and control

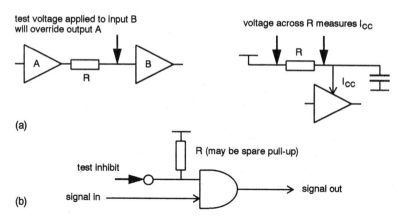

(a)

(b)

Figure 9.5 Use of an extra test resistor

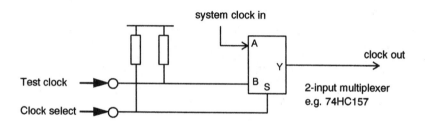

Figure 9.6 Test clock selection with data multiplexer

signals continuously in a predictable manner, so that the test equipment can monitor them for the correct function. The test software operation depends of course on the core functions of the microprocessor, and its bus and control signal interconnections, being fault-free.

More complex digital systems cannot easily be tested or, more to the point, debugged, with the techniques described so far. Considerable effort goes into researching sophisticated test methods for such systems. If you are designing large digital boards you will need to be familiar with these methods and design them in from the start, since they consume a large amount of circuit overhead in order to function.

9.4 Reliability

The reliability of electronic equipment can to some extent be quantified, and a separate discipline of reliability engineering has grown up to address it. This section will serve as an introduction to the subject for those designers who are not fortunate enough to have a reliability engineering department at their disposal.

9.4.1 Definitions

Reliability, itself, has a strictly defined meaning. This can be stated as "the probability that a system will operate without failure for a specified period, subject to specified environmental conditions". Thus it can be quoted as a single number, such as 90%, but this is subject to three qualifications:

- agreement as to what constitutes a "failure". Many systems may "fail" without becoming totally useless in the process.

- a specified operating lifetime. No equipment will operate forever; reliability must refer to the reasonably foreseen operating life of the equipment, or to some other agreed period. The age of the equipment, which may well affect failure rate, is not a factor in the reliability specification.

- agreement upon environmental conditions. Temperature, moisture, corrosive atmospheres, dust, vibration, shock, supply and electromagnetic disturbances all have an effect on equipment operation and reliability is meaningless if these are not quoted.

If you offer or purchase equipment whose reliability is quoted for one set of conditions and it is used under another set, you will not be able to extrapolate the reliability figure to the new conditions unless you know the behaviour of those parameters which affect it.

Mean time between failures

For most of the life of a piece of electronic equipment, its failure rate (denoted by λ) is constant. In the early stages of operation it could be high and decrease as weak components fail quickly and are replaced; late in its life components may begin to "wear out" or corrosion may take its toll, and the failure rate may start to rise again. The reciprocal of failure rate during the constant period is known as the mean time between failures (MTBF). This is generally quoted in hours, whilst failure rate is quoted in faults per hour. For instance, an MTBF of 10,000 hours is equivalent to a failure rate of 0.0001 faults per hour or 100 faults per 10^6 hours. MTBF has the advantage that it does not depend on the operating period, and is therefore more convenient to use than reliability.

Mean time to failure

MTBF measures equipment reliability on the assumption that it is repaired on each failure and put back into service. For components which are not repairable, their reliability is quoted as mean time to failure (MTTF). This can be calculated statistically by observing a sample from a batch of components and recording each one's working life, a procedure known as life testing. The MTTF for this batch is then given by the mean of the lifetimes.

Availability

System users need to know for what proportion of time their system will be available to them. This figure is given by the ratio of "up-time", during which the system is switched on and working, to total operating time. The difference between the two is the "down-time" during which the system is faulty and/or under repair. Thus

$$A = [U/(U + D)]$$

Availability can also be related to the MTBF figure and the mean time to repair (MTTR) figure by

$$A = [MTBF/(MTBF + MTTR)]$$

The availability of a particular system can be monitored by logging its operating data, and this can be used to validate calculated MTBF and MTTR figures. It can also be interpreted as a probability that at any given instant the system will be found to be working.

9.4.2   The cost of reliability

Reliability does not come for free. Design and development costs escalate as more effort is put into assuring it, and component costs increase if high performance is required of them. For instance, it would be quite possible to improve the reliability of, say, an audio power amplifier by using massively over-rated output transistors, but these would add considerably to the selling cost of the amplifier. On the other hand, if the selling cost were reduced by specifying under-rated transistors, the users would find their total operating costs mounting since the output transistors would have to be replaced more frequently. Thus there is a general trend of decreasing operating or "life-cycle" costs and increasing unit costs, as the designed-in reliability of a given system increases. This leads to the notion of an "optimum" reliability figure in terms of cost for a system. Figure 9.7 illustrates this trend. The criterion of good design is then to approach this optimum as closely as possible.

Of course, this argument only applies when the cost of unreliability is measured in strictly economic terms. Safety-critical systems, such as nuclear or chemical process plant controllers, railway signalling or flight-critical avionics, must instead meet a defined reliability standard and the design criterion then becomes one of assuring this level of reliability, with cost being a secondary factor.

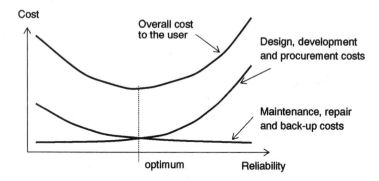

Figure 9.7 Reliability versus cost

9.4.3   Design for reliability

The goal of any circuit designer is to reduce the failure rate of their design to the minimum achievable within cost constraints. The factors which help in meeting this goal are

- use effective thermal management to minimise temperature rise
- de-rate susceptible components as far as possible
- specify high reliability or quality assured components
- specify stress screening or burn-in tests

- keep circuits simple, use the minimum number of components
- use redundancy techniques at the component level

Temperature

High temperature is the biggest enemy of all electronic components and measures to keep it down are vital. Temperature rise accelerates component breakdown because chemical reactions occurring within the component, which govern bond fractures, growth of contamination or other processes, have an increased rate of reaction at higher temperature. The rate of reaction is determined by the Arrhenius equation,

$$\lambda \quad = \quad K \cdot exp\ (-E/kT)$$

where λ gives a measure of failure rate
         K is a constant depending on the component type
         E is the reaction's activation energy
         k is Boltzmann's constant, $1.38 \cdot 10^{-23}$ J K^{-1}
         T is absolute temperature

Many reactions have activation energies around 0.5eV which results in an approximate doubling of λ with every 10°C rise in temperature, and this is a useful rule of thumb to apply for the decrease in reliability versus temperature of typical electronic equipment with many components. Some reactions have higher activation energy, which give a faster increase of λ with temperature.

Thermal management itself is covered in section 9.5.

De-rating

There is a very significant improvement to be gained by operating a component well within its nominal rating. For most components this means either its voltage or power rating, or both.

Take capacitors as an example. Conventionally, you will determine the maximum dc bias voltage a capacitor will have to withstand under worst-case conditions and then select the next highest rating. Over-specifying the voltage rating may result in a larger and more costly component.

However, capacitor life tests show that as the maximum working voltage is approached, the failure rate increases as the fifth power of the voltage. Therefore, if you run the capacitor at half its rated voltage you will observe a failure rate 32 times lower than if it is run at full rated voltage. Given that a capacitor of double the required rating will not be as much as double the size, weight or cost, except at the extremes of range, the improvement in reliability is well worth having.

In many cases there is no difficulty in using a de-rated capacitor; small film capacitors for instance are rated at a minimum of 50 or 100V and are frequently used in 5V circuits. Electrolytics on the other hand are more likely to be run near their rating. These capacitors already have a much higher failure rate than other types because of their construction – the electrolyte has a tendency to "dry out", especially at high temperatures – and so you will achieve significant improvement, albeit at higher cost, if you heavily de-rate them.

De-rating the power dissipation of resistors reduces their internal temperature and therefore their failure rate. In low voltage circuits there is no need to check power rating for any except low value parts; if for instance you use 0.4 watt metal film resistors in a circuit with a maximum supply of 10V you can be sure that all resistors over 500Ω will be derated by at least a factor of 2, which is normally enough.

Semiconductor devices are normally rated for power, current and voltage, and de-rating on all of these will improve failure rate. The most important are power dissipation, which is closely linked to junction temperature rise and cooling provision, and operating voltage, especially in the presence of possible transient over-voltages.

High reliability components

Component manufacturers' reputations are seriously affected by the perceived reliability or otherwise of their product, so most will go to considerable effort not to ship defective parts. However, the cost of detecting and replacing a faulty part rises by an order of magnitude at each stage of the production process, starting at goods inwards inspection, proceeding through board assembly, test and final assembly, and ending up with field repair. You may therefore decide (even in the absence of mandatory procurement requirements on the part of your customer) that it is worth spending extra to specify and purchase parts with a guaranteed reliability specification at the "front end" of production.

BS9000

Initially it was military requirements, where reliability was more important than cost, that drove forward schemes for assessed quality components. More recently some commercial customers have also found it necessary to specify such components. The need for a common standard of assessed quality is met in the UK by the BS9000 scheme. This is now harmonized with the European CECC[†] quality standards. BS9000 itself describes the scheme and specifies the way in which parts are approved. Its sister documents BS9003 and BS9005 lay down manufacturing requirements and approval procedures.

Generic specifications are found in the BS9XXX or CECC series for all types of component which are covered by the scheme. These specify physical, mechanical and electrical properties, and lay down test requirements. Individual component specifications are not found under the scheme.

US MIL standards

The US military take a somewhat different approach to reliability. For semiconductors, the standard MIL-STD-883 sets out uniform testing methods and procedures, including environmental, physical and electrical tests. It also establishes three distinct levels of product assurance for applications requiring different levels of reliability. These are, in order of decreasing severity, Class S for critical applications, such as space-borne equipment; Class B for less critical purposes such as airborne or ground systems; and Class C for easily repairable ground applications. The same part number can be procured to any of these specifications, when it will have undergone testing and stress screening to different degrees of rigour. Class B is the most common.

MIL-STD-883 is not product-specific. The MIL-M-38510 standardization program on the other hand requires that all suppliers of qualified devices conduct and implement specific tests and manufacturing controls on each part. It establishes the procedures which a manufacturer must follow to have his products listed on the Qualified Parts List (QPL). Detailed "slash sheets" define the performance parameters and electrical screens for each individual device. This is then referenced by its slash sheet number; for example a qualified 4011B CMOS device would be called up as a JM38510/05051.

MIL-M-38510 in theory allows designers to specify a device of assured reliability

---

† CENELEC Electronic Components Committee

from a number of different suppliers. In practice, it takes some time for a new part to make it onto the QPL, and this means that designers are unable to use state-of-the-art components in products that mandate MIL-M-38510 sourcing.

Stress screening and burn-in

The above specifications all include some degree of stress screening. This phrase refers to testing the components under some type of stress, typically at elevated temperature, under vibration or humidity and with maximum rated voltage applied, for a given period. This practice is also called "burning in". The principle is that weak components will fail early in their life and the failures can be accelerated by operating them under stress. These can then be weeded out before the parts are shipped from the manufacturer. A typical test (specified in MIL-STD-883B) would be 160 hours at 125°C. Another common test is a repeated temperature cycle between the extremes of the permitted temperature range, which exposes failures due to poor bonding or other mechanical faults.

Such stress screening can be applied to any component, not just semiconductors, and also to entire assemblies. If you are unsure of the probable quality of early production output of a new design, specifying stress screening on the first few batches is a good way to discover any recurrent production faults before they are passed out to the customer. It is expensive in time, equipment and inventory, and should not be used as a crutch to compensate for poor production practices. It should only be employed as standard if the customer is willing to pay for it.

Simplicity

The failure rate of an electronic assembly is roughly equal to the sum of the failure rates of all its components. This assumes that a failure in any one component causes the failure of the whole assembly. This is not necessarily a valid assumption, but to assume otherwise you would have to work out the assembly's failure modes for each component failure and for combinations of failures, which is not practical unless your customer is prepared to pay for a great deal of development work.

If the assumption holds, then reducing the number of components will reduce the overall failure rate. This illustrates a very important principle in circuit design: *the highest reliability comes from the simplest circuits*. Apply Occam's razor ("entities should not be multiplied beyond necessity") and cut down the number of components to a minimum.

Redundancy

Redundancy is employed at the system level by connecting the outputs of two or more sub-systems together such that if one fails, the others will continue to keep the system working. A typical example might be several power supplies, each connected to the same power distribution rail (via isolating diodes) and each capable of supplying the full load. If the reliability of the interconnection is neglected, the probability of all supplies failing simultaneously is the product of the probabilities of failure of each supply on its own, assuming that a common mode failure (such as the mains supply to all units going off) is ruled out.

The principle can also be applied at the component level. If the probability of a single component failing is too high then redundant components can be placed in parallel or series with it, depending on the required failure mode. This technique is mandatory in certain fields, such as intrinsically safe instrumentation. Figure 9.8 illustrates redundant Zener clamping. The Zeners prevent the voltage across their terminals from rising to an unsafe value in the event of a fault voltage being applied at the input to the barrier. One Zener alone would not offer the required level of reliability, so two further ones are placed in parallel, so that even with an open-circuit failure of two out of the three, the clamping action is maintained. The interconnections between the Zeners must be solid enough not to materially affect the reliability of the combination.

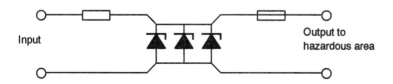

Figure 9.8 The intrinsically safe Zener barrier

Some provision must normally be made for detecting and indicating a failed component or subsystem so that it can be repaired or replaced. Otherwise, once a redundant part has failed, the overall reliability of the system is severely reduced.

9.4.4   The value of MTBF figures

The mean-time-between-failure figure as defined in section 9.4.1 can be calculated before the equipment is put into production by summing the failure rates of individual components to give an overall failure rate for the whole equipment. As discussed earlier, this assumes that a fault in any one component causes the failure of the whole assembly. This method presupposes adequate data on the expected failure rates of all components that will be used in the equipment.

Such sources of failure rate data for established component types are available. The most widely used is MIL-HDBK-217, now in its fifth revision, published by the US Department of Defense. This handbook lists failure rate models and tables for a wide variety of components, based on observed failure measurements. A failure rate for each component can be derived from its operating and environmental conditions, de-rating factor and method of construction or packaging. A further factor that is included for integrated circuits is their complexity and pinout. Another source of failure rate data, somewhat less comprehensive but widely used for telecommunications applications, is British Telecom's handbook HRD4.

The disadvantage with using such data is that it cannot be up-to-date. Proper failure rate data takes years to accumulate, and so data extracted from these tables for modern components will not be accurate. This is especially true for integrated circuits. Generally, figures based on obsolete failure rate data will tend to be pessimistic, since the trend of component reliability is to improve.

Calculations of failure rates at component level are tedious, since operating conditions for each component, notably voltage and power dissipation, must be a part of the calculation in order to arrive at an accurate value. They do lend themselves to computer derivation, and software packages for reliability prediction are readily available. Since in many cases such operating conditions are highly variable, it is arguable that you will not obtain much more than an order-of-magnitude estimate of the true figure anyway.

A published MTBF figure does not tell you how long the unit will actually last, and it does not indicate how well the unit will perform in the field under different environmental and operating conditions. Such figures are mainly used by the marketing department to make the specification more attractive. But MTBF prediction *is* valuable for two purposes:

- for the designer, it gives an indication of where reliability improvements can most usefully be made. For instance, if as is often the case the electrolytic capacitors turn out to make the highest contribution to overall failure rate, you can easily evaluate the options available to you in terms of de-rating or adding redundant components. You need not waste effort on optimising those components which have little effect overall.

- for the service engineer, it gives an idea of which components are likely to have failed if a breakdown occurs. This can be valuable in reducing servicing and repair time.

9.4.5 Design faults

Before leaving the subject of reliability design, we should briefly mention a very real problem, which is the fallibility of the designers themselves. There is no point in specifying highly reliable components or applying all manner of stress screening tests or redundancy techniques if the circuit is going to fail because it has been wrongly designed. Design faults can be due to inexperience, inattention or incompetence on the part of the designer, or simply because the project timescale was too short to allow the necessary cross-checking. Computer-aided design techniques and simulators can reduce the risk but they cannot eliminate the potential for human error completely.

The design review

An effective and relatively painless way of guarding against design faults is for your product development department to instigate a system of frequent design reviews. In these, a given designer's circuit is subjected to a peer critique in order to probe for flaws which might not be apparent to the circuit's originator. The critique can check that the basic circuit concept is sound and cost-effective, that all component tolerances have been accounted for, that parts will not be operated outside their ratings, and so on. The depth of the review is determined by the resources that are available within the group; the reviewers should preferably have no connection with the project being reviewed, so that they are able to question underlying and unstated assumptions. Naturally, the effectiveness of such a system depends on the resources a company is prepared to devote to it, and it also depends on the willingness of the designer to undergo a review. Personality clashes tend to surface on these occasions. Each designer develops pet techniques and idiosyncrasies during their career, and provided these are not actually wrong they should not attract criticism. Nevertheless, design reviews are valuable for testing the strength of a design before it gets to the stage where the cost of mistakes becomes significant.

9.5 Thermal management

It is in the nature of electronic components to dissipate power while they are operating. Any flow of current through a non-ideal component will develop some power within that component, which in turn causes a rise in temperature. The rise may be no more than a small fraction of a degree Celsius when less than a milliwatt is dissipated, extending to several tens or even hundreds of degrees when the dissipation is measured in watts. Since excess temperature kills components, some way must be found to maintain the component operating temperature at a reasonable level. This is known as thermal management.

9.5.1 Calculating thermal resistance

The attractiveness of thermal analysis to electronics designers is that it can easily be understood by means of an electrical analogue. The flow of heat can be visualized as emanating from the component which is dissipating power, passing through some form of thermal interface and out to the environment, which is assumed to have a constant ambient temperature T_A and infinite ability to sink heat. Heat transfer through the thermal interface is accomplished by one or more of three mechanisms: conduction, convection and radiation. For most purposes only the first two are important in electronic applications.

Figure 9.9 shows the generalized model and its electrical analogue. The model can be analysed using conventional circuit theory and yields the following equation for the temperature at the heat source:

$$T = P_D \cdot R_\theta + T_A$$

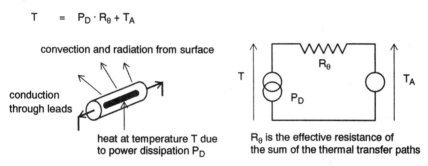

convection and radiation from surface

conduction through leads

heat at temperature T due to power dissipation P_D

R_θ is the effective resistance of the sum of the thermal transfer paths

Figure 9.9 Heat transfer from hot component to ambient

This temperature is the critical factor for electronic design purposes, since it determines the reliability of the component. Reducing any of P_D, R_θ or T_A will minimise T. Ambient temperature is not normally under your control but is instead a specification parameter (but see section 9.5.4). Since you are normally attempting to manage a given power dissipation, the only parameter which you are free to modify is the thermal resistance R_θ. This is achieved by heatsinking.

There are more general ways of analysing heat flow and temperature rise, using thermal conductivity and the area involved in the heat transfer. However, component manufacturers normally offer data in terms of thermal resistance and maximum permitted temperature, so it is easiest to perform the calculations in these terms.

Partitioning the heat path

When you have data on the component's thermal resistance directly to ambient, and your mounting method is simple, then the basic model of Figure 9.9 is adequate. For components which require more sophisticated mounting and whose heat transfer paths are more complicated, you can extend the model easily. The most common application is the power semiconductor mounted via an insulating washer to a heatsink (Figure 9.10(a)).

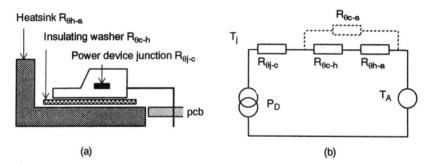

Figure 9.10 Heat transfer for a power device on a heatsink

The equivalent electrical model is shown in Figure 9.10(b). Here, T_j is the junction temperature and $R_{\theta j\text{-}c}$ represents the thermal resistance from junction to case of the device. All manufacturers of power devices will include $R_{\theta j\text{-}c}$ in their data sheets and it can often be found in low-power data as well. Sometimes it is disguised as a power de-rating figure, expressed in mW/°C. The maximum allowable value of T_j is published in the maximum ratings section of each data sheet.

$R_{\theta c\text{-}h}$ and $R_{\theta h\text{-}a}$ are the thermal resistances of the interface between the case and the heatsink, and of the heatsink to ambient, respectively. $R_{\theta c\text{-}a}$ represents the thermal resistance due to convection directly from case to ambient, and can be neglected unless you are using a small heatsink.

An example should help to make the calculation clear.

An IRF150 power MOSFET dissipates a maximum of 50W steady-state. It is mounted on a heatsink with a specified thermal resistance of 0.5°C per watt, via an insulating pad with a thermal resistance of 0.3°C per watt. The maximum ambient temperature is 70°C. What will be the maximum junction temperature?

From the above conditions, $R_{\theta c\text{-}h} + R_{\theta h\text{-}a} = 0.8$°C/W. The IRF150 data quotes a junction-to-case thermal resistance ($R_{\theta j\text{-}c}$) of 0.83°C/W.

So the junction temperature

$$T_j \quad = \quad 50 \cdot [0.83 + 0.8] + 70 \quad\quad = \quad 151.5°C$$

This is just over the maximum permitted junction temperature of 150°C so reliability is marginal and you need a bigger heatsink. However, we have neglected the case-to-ambient thermal resistance, quoted at 30°C/W. This is in parallel with $R_{\theta c\text{-}h} + R_{\theta h\text{-}a}$. If it is included, the calculation becomes

$$T_j \quad = \quad 50 \cdot [0.83 + 0.78] + 70 \quad\quad = \quad 150.5°C$$

Clearly, nothing significant is gained by taking it into account.

This example illustrates a common misconception about power ratings. The IRF150 is rated at 150W dissipation, yet even with a fairly massive heatsink (0.5°C/W will require a heatsink area of around 80 square inches) it cannot safely dissipate more than 50W at an ambient of 70°C. The fact is that the rating is specified *at 25°C case temperature*; higher case temperatures require de-rating because of the thermal resistance from junction to case. You will not be able to maintain 25°C at the case under any practical application conditions. Power device manufacturers publish de-rating curves in their data sheets: rely on these rather than the absolute maximum power rating on the front of the specification.

Transient thermal characteristics

In applications where the power dissipated in the device consists of low duty cycle pulses, the instantaneous or peak junction temperature may be the limiting condition rather than the average temperature. In this case you need to consult curves for transient thermal resistance. These curves are normally provided by power semiconductor manufacturers in the form of a correction factor that multiplies $R_{\theta j\text{-}c}$ to allow for the duty cycle of the power dissipation. Figure 9.11 shows a family of such curves for the IRF150. Because the period for most pulsed applications is much shorter than the heatsink's thermal time constant, the value of $R_{\theta h\text{-}a}$ can be multiplied directly by the duty cycle. Then the junction temperature can now be calculated from

$$T_j \quad = \quad P_{Dmax} \cdot [K \cdot R_{\theta j\text{-}c} + R_{\theta c\text{-}h} + \delta \cdot R_{\theta h\text{-}a}] + T_A$$

where δ is the duty cycle and K is derived from curves as in Figure 9.11 for a particular value of δ.

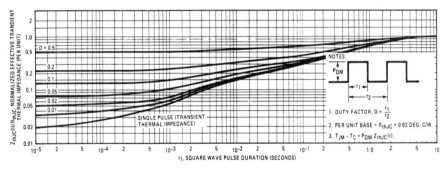

Figure 9.11 Transient thermal impedance curves for the IRF150
Source: International Rectifier

Some applications, notably RF amplifiers or switches driving highly inductive loads, may create severe current crowding conditions on the die which invalidate methods based on thermal resistance or transient thermal impedance. Safe operating areas and di/dt limits must be observed in these cases.

9.5.2 Heatsinks

As the previous section implied, the purpose of a heatsink is to provide a low thermal resistance path between the heat source and the ambient. The heatsink does not itself sink the heat, except temporarily. In most cases the ambient sink will be air, though not invariably: this author recalls one somewhat tongue-in-cheek design for a 1kW rated audio amplifier which suggested bolting the power transistors to a central heating radiator with continuous water cooling! Some designs with a very high power density need to adopt such measures to ensure adequate heat removal.

A wide range of proprietary heatsinks are available from many manufacturers. Several types are pre-drilled to accept common power device packages. All are characterised to give a specification figure for thermal resistance, usually quoted in free air with fins vertical. Unless your requirements are either very specialised or very high volume, it is not worth designing your own heatsink, especially as you will have to go through the effort of testing its thermal characteristics yourself. Custom heatsink design is covered in the application notes of several power device manufacturers.

A heatsink transfers heat to ambient air primarily by convection, and to a lesser degree by radiation. Its efficiency at doing so is directly related to the surface area in contact with the convective medium. Thus heatsink construction seeks to maximize surface area for a given volume and weight; hence the preponderance of finned designs. Orientation of the fins is important because convection requires air to move past the surface and become heated as it does so. As air is heated it rises. Therefore the best convective efficiency is obtained by orienting the fins vertically to obtain maximum air flow across them; horizontal mounting reduces the efficiency by around 70%.

The most common material for heatsinks is black anodised aluminium. Aluminium offers a good balance between cost, weight and thermal conductivity. Black anodising provides an attractive and durable surface finish and also improves radiative efficiency by 10–15 times over polished aluminium. Copper can be used as a heatsink material when the optimum thermal conductivity is required, but it is heavier and more expensive.

Forced air cooling

Convective heat loss from a heatsink can be enhanced by forcing the convective medium across its surface. The design of forced air cooled heatsinks is best done empirically. Another common use of forced air cooling is ventilation of a closed equipment cabinet by a fan. The capacity of the fan is quoted as the volumetric flow rate in cubic feet per minute (CFM) or cubic metres per hour (1 CFM = 1.7m^3/hr). The volumetric flow rate required to limit the internal temperature rise of an enclosure in which P_D watts of heat is dissipated to $\theta°C$ above ambient is

Flow rate   = $3600 \cdot P_D / (\rho \cdot c \cdot \theta)$ m^3/hr

where ρ is the density of the medium
             (air at 30°C and atmospheric pressure is 1.3kg/m^3)
             c is the specific heat capacity of the medium
             (air at 30°C is around 1000J kg^{-1} °C^{-1})

Fan performance is shown as volumetric flow rate versus pressure drop across the fan. The pressure differential is a function of the total resistance to airflow through the enclosure, presented by obstacles such as air filters, louvres, and pcbs. You generally need to derive pressure differential empirically for any design with a non-trivial air flow path.

9.5.3   Power semiconductor mounting

The way in which a power device package is mounted to its heatsink affects both actual heat transfer efficiency and long-term reliability. Faulty mounting of metal packaged devices mainly causes unnecessarily high junction temperature, shortening device lifetime. Plastic packages (such as the common TO-220 outline) are much more susceptible to mechanical damage, which allows moisture into the case and can even crack the semiconductor die.

The factors which you should consider when deciding on a mounting method are summarized in Figure 9.12 for a typical plastic-packaged device.

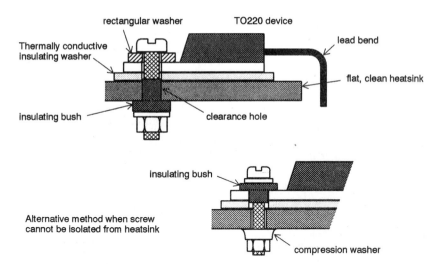

Figure 9.12 Screw mounting methods for a power device

Heatsink surface preparation

The heatsink should have a flatness and finish comparable to that of the device package. The higher the power dissipation, the more attention needs to be paid to surface finish. A finish of 50 – 60 microinches is adequate for most purposes. Surface flatness, which is the deviation in surface height across the device mounting area, should be less than 4 mils (0.004") per inch.

The mounting hole(s) should only be large enough to allow clearance of the fastener, plus insulating bush if one is fitted. Too large a hole, if the screw is torqued too tightly, will cause the mounting tab to deform into the hole. This runs the risk of cracking the die, as well as lifting the major part of the package which is directly under the die off the heatsink in cantilever fashion, seriously affecting thermal transfer impedance. Chamfers on the hole must be avoided for the same reason, but de-burring is essential to avoid puncturing insulation material and to maintain good thermal contact. The surface should be cleaned of dust, grease and swarf immediately before assembly.

Lead bend

Bending the leads of any semiconductor package stresses the lead interface and can result in cracking and consequent unreliability. If possible, mount your devices upright on the pcb so that lead bending is unnecessary. Plastic packaged devices (TO220, TO126 etc) can have their leads bent, provided that

- the minimum distance between the plastic body and the bend is 4mm
- the minimum bend radius is 2mm
- maximum bend angle is no greater than 90°
- leads are not repeatedly bent at the same point
- no axial strain is applied to the leads, relative to each other or the package

Use round-nosed pliers or a proper lead forming jig to ensure that these conditions are met. Metal cased devices must not have their leads bent, as this is almost certain to damage the glass seal.

When the device is inserted into the board, the leads should always be soldered after the mechanical fastening has been made and tightened. Some manufacturing departments prefer not to run cadmium plated screws through a solder bath because it contaminates the solder, and they may decide to put the screws in after the mass soldering stage. Do not allow this: insist on hand soldering or use different screws.

The insulating washer

In most devices, the heat transfer tab or case is connected directly to one of the device terminals, and this raises the problem of isolating the case. The best solution from the point of view of thermal resistance is to isolate the entire heatsink rather than use any insulating device between the package and the heatsink. This is often not possible, for EMI or safety reasons, because the chassis serves as the heatsink, or because several devices share the same heatsink. Some devices are now available in fully-isolated packages, but if you aren't using one of these you will have to incorporate an insulating washer under the package.

Insulating washers for all standard packages are available in many different materials: polyimide film, mica, hard anodised aluminium and reinforced silicone rubber are the most popular. The first three of these require the use of a thermally conductive grease between the mating surfaces, to fill the minor voids which exist and which would otherwise increase the thermal resistance across the interface. This is messy and increases the variability and cost of the production stage. If excess grease is left around the device, it may accumulate dust and swarf and lead to insulation breakdown across the interface. Silicone rubber, being somewhat conformal under pressure, can be used dry and some types will outperform mica and grease.

Table 9.3 shows the different interface thermal resistances ($R_{\theta c-h}$) that may be expected. Note that when thermally conductive grease is not used, wide variations in thermal resistance will be encountered because of differences in surface finish. The mounting hole(s) in the washer should be no larger than the device's holes, otherwise flashover to the exposed metal is likely.

Package type	Interface thermal resistance °C/W						
	Metal-to-metal		With insulator				
	Dry	Greased	2-mil mica		2-mil polyimide		Dry 6-mil silicone rubber
			Dry	Greased	Dry	Greased	
Metal, flanged							
TO204AA (TO3)	0.5	0.1	1.2	0.35	1.5	0.55	0.4 - 0.6
TO213AA (TO66)	1.5	0.5	2.3	0.9	-	-	-
Plastic							
TO126	2.0	1.3	4.3	3.3	-	-	-
TO220AB	1.2	1.0	3.4	1.6	4.5	2.2	1.3

Table 9.3 Interface thermal resistances for various mounting methods

Mounting hardware

A combination of machine screws, compression washers, flat washers and nuts is satisfactory for any type of package that has mounting holes. Check the specified mounting hole tolerances carefully; there is a surprisingly wide variation in hole dimensions for the same nominal package type across different manufacturers. A flat, preferably rectangular (in the case of plastic packages) washer under the screw head is vital to give a properly distributed pressure, otherwise cracking of the package is likely. A conical compression washer is a very useful device for ensuring that the correct torque is applied. This applies a constant pressure over a wide range of physical deflection, and allows proper assembly by semi-skilled operators without using a torque wrench or driver. Tightening the fasteners to the correct torque is very important; too little torque results in a high thermal impedance and long term unreliability due to over-temperature, while too much can overstress the package and result in long term unreliability due to package failure.

A fast, economical and effective alternative is a mounting clip. When only a few watts are being dissipated, you can use board mounting or free standing dissipators with an integral clip. A separate clip can be used for larger heatsinks and higher powers. The clip must be matched to the package and heatsink thickness to obtain the proper pressure. It can actually offer a lower thermal resistance than other methods for plastic packages, because it can be designed to bear directly down on top of the plastic over the die, rather than concentrating the mounting pressure at the hole in the tab.

When screw-mounting a device which has to be isolated from the heatsink, you need to use an insulating bush either in the device tab or the heatsink. The preferred method is to put the bush in the heatsink, and use large flat washers to distribute the mounting force over the package. You can also use larger screws this way. The bush material should be of a type that will not flow or creep under compression; glass-filled nylon or polycarbonate are acceptable, but unfilled nylon should be avoided. The bush should be long enough to overlap between the transistor and the heatsink, in order to prevent flashover between the two exposed metal surfaces.

9.5.4   Placement and layout

If you are only concerned with designing circuits that run at slow speeds with CMOS logic and draw no more than a few milliamps, then thermal layout considerations will not interest you. As soon as dissipation raises the temperature of your components more than a few tens of degrees above ambient, it pays to look at your equipment and pcb layout in terms of heat transfer. As was shown in section 9.4.3, this will ultimately reflect in the reliability of the equipment.

Some practices that will improve thermal performance are

- mount pcbs vertically rather than horizontally. This is standard in card cages and similar equipment practice, and it allows a much freer convective airflow over the components. If you are going to do this, do not then block off the airflow by putting solid metal screens above or below the boards; use punched, louvred or mesh screens.

- put hot components near the edge of the board, to encourage a good airflow around them and their heatsinks. If the board will be vertically mounted, put them at the top of the board.

- keep hot components as far away as possible from sensitive devices such as precision op-amps or high failure rate parts such as electrolytic capacitors.

Put them above such components if the board is vertical.

- heatsinks perform best in low ambient temperatures. If you are using a heatsink within an enclosure without forced air cooling, remember to allow for the steady-state temperature rise inside the enclosure. However, don't position a heatsink near to the air inlet, as it will heat the air that is circulating through the rest of the enclosure; put it near the outlet. Don't obstruct the airflow over a heatsink.

- if you have a high heat density, for example a board full of high speed logic devices, consider using a thermally conductive ladder fixed on the board and in contact with the IC packages, brought out to the edge of the board and bonded to an external heatsink. PCB laminates themselves have a low thermal conductivity.

- if you have to use a case with no ventilation, for environmental or safety reasons, remember that cooling of the internal components will be by three stages of convection rather than one: from the component to the inside air, from the inside air to the case, and from the case to the outside. Each of these will be inefficient, compared to conductive heat transfer obtained by mounting hot components directly onto the case. But if you take this latter course, check that the outside case temperature will not rise to dangerously high levels.

Appendix

Standards

Standards are indispensable to manufacturing industry. Not only do they allow interchangeability or interoperability between different manufacturers' products, but they also represent a distillation of knowledge about practical aspects of technology – how to make measurements, what tests to use, what dimensions to specify and so on. Each standard is the result of considerable work on the part of a collection of experts in that particular field and is therefore authoritative; notwithstanding which, any standard in a fast-changing area will be subject to revision and amendment as the technology progresses.

This appendix lists a few of the more relevant British and international standards, both those which have been referenced in the text of this book and some which the author feels to be of particular interest. The catalogues of the various standards bodies, updated yearly, give a full list of the available and current standards and are essential for the library of any development department. Although only BS and IEC publications are mentioned here, for the sake of brevity, there are of course many other standards sources which you may need to consult for a particular application.

British standards

These are available from

British Standards Institution
Sales Department
Linford Wood
Milton Keynes
MK14 6LE
UK
Telex 825777 BSIMK G; Telefax 0908 320856

Some BS standards are related to, equivalent to or identical to other European or international standards. This is indicated where appropriate.

BS no.	Related, equivalent or identical standards
BS88	IEC269

Cartridge fuses for voltages up to and including 1000V ac and 1500V dc

BS397	IEC86

Primary batteries

BS415	IEC65

Specification for safety requirements for mains-operated electronic and related apparatus for household and similar general use

BS613

Specification for components and filter units for electromagnetic interference suppression

BS727   **CISPR16**

Specification for radio-interference measuring apparatus

BS800   **EN55014, CISPR14**

Specification for limits and methods of measurement of radio interference characteristics of household electrical appliances, portable tools and similar electrical apparatus

BS905   **CISPR13, CISPR20**

Sound and television broadcast receivers and associated equipment: electromagnetic compatibility

BS1852   **IEC62**

Specification for marking codes for resistors and capacitors

BS2011   **IEC68**

Basic environmental testing procedures

BS2316   **IEC96**

Specification for radio-frequency cables

BS2488   **IEC63**

Schedule of preferred numbers for resistors and capacitors

BS2754   **IEC536**

Memorandum: construction of electrical equipment for protection against electric shock

BS3939   **IEC617**

Guide for graphical symbols for electrical power, telecommunications and electronics diagrams

BS4109

Specification for copper for electrical purposes. Wire for general electrical purposes and for insulated cables and flexible cords

BS4265   **IEC127**

Specification for cartridge fuse links for miniature fuses

BS4808   **IEC189**

Specification for LF cables and wires with PVC insulation and PVC sheath for telecommunications

BS5406   **EN60555, IEC555**

Disturbances in supply systems caused by household appliances and similar electrical equipment

BS5490   **IEC529**

Specification for classification of degrees of protection provided by enclosures

BS5783

Code of practice for handling of electrostatic sensitive devices

BS5932   **IEC285**

Specification for sealed nickel-cadmium cylindrical rechargeable single cells

BS6221   **IEC326**

Printed wiring boards

BS6500   **IEC227, IEC245**

Specification for insulated flexible cords and cables

BS6527   **EN55022, CISPR22**

Specification for limits and methods of measurement of radio interference characteristics of information technology equipment

BS6555   **IEC431**

Specification for dimensions of square cores (RM-cores) made of magnetic oxides and associated parts

BS6667   **IEC801**

Electromagnetic compatibility for industrial process measurement and control equipment

BS6811   **IEC182, IEC851**

Winding wires

BS9000   **CECC**

General requirements for a system for electronic components of assessed quality

BS9930   **IEC384**

Harmonized system of quality assessment for electronic components. Fixed capacitors for use in electronic equipment

BS9940   **IEC115**

Harmonized system of quality assessment for electronic components. Fixed resistors for use in electronic equipment

IEC standards

The IEC (International Electrotechnical Commission) is responsible for international standardisation in the electrical and electronics fields. It is composed of member National Committees. IEC publications are available from the BSI, address as before, or from

IEC Central Office
1 Rue de Varembé
1211 Geneva 20
Switzerland

Most IEC publications have related British standards (see list above). A few which do not are listed below.

IEC257

Fuse-holders for miniature cartridge fuse-links

IEC479

Effects of current passing through the human body

IEC509

Sealed nickel-cadmium button rechargeable single cells

IEC647

Dimensions for magnetic oxide cores intended for use in power supplies (EC-cores)

Bibliography

When writing a book of this nature, one is invariably indebted to a great many sources of information, which are often used in day-to-day work as a circuit designer. This reading list pulls together some of the references I have found most useful over the years, and/or which have been drawn on extensively for some of the chapters. Most are manufacturers' data books.

On batteries:

Guide for Designers, Duracell Europe, 1985

Short applications guide and data book covering most primary cell systems.

On digital design:

High-speed CMOS Designer's Guide and Applications Handbook, Mullard, 1986 (now Philips Components)

Most manufacturers produce extensive applications literature on their standard logic families. This book is the most comprehensive guide I have found on HCMOS.

Transmission line effects in PCB applications, Motorola application note AN1051, 1990

Presents design analysis and examples for pcb transmission line effects in digital circuits. Includes Bergeron diagram analysis.

On discrete devices:

Philips Components Data Book 1 Part 1b, **Low-frequency power transistors and modules,** 1989

Contains information on transistor ratings, safe operating area and letter symbols

Thyristor & Triac Theory and Applications, in Motorola Thyristor Databook, 1985

Nine chapters cover the title's scope comprehensively and in detail.

Applications, in Low Power Discretes Data Book, Siliconix 1989

Covers all aspects of JFET applications.

Power MOSFET Application Data, in HEXFET Databook, International Rectifier 1982

Covers all aspects of power MOSFETs, and is equally applicable to devices from other manufacturers.

Mounting Considerations for Power Semiconductors, Bill Roehr, Motorola Applications Note AN1040, 1988

Useful summary of techniques and considerations for good thermal and mechanical design practice

On grounding and EMC:

Grounding and Shielding Techniques in Instrumentation, Ralph Morrison, Wiley-Interscience 1986

A rare guide to the fundamentals of grounding and shielding practice. Terse style but very useful.

Ground: a path for current flow, Henry W. Ott, IEEE International Symposium on EMC, San Diego 9-11 October 1979, pp 167-170

Unusually lucid for a learned paper, this presents the rationale for considering ground as a current path.

Noise Reduction Techniques in Electronic Systems, Henry W. Ott, Wiley Interscience, 1988

A very useful guide to EMC techniques and effects, especially in digital systems

On op-amps:

Intuitive IC Op Amps, Thomas M Frederiksen, National Semiconductor 1984

This is undoubtedly the most useful book on op-amps there is. Frederiksen designed the LM324 amongst other things.

Op Amp Orientation, in Analog Devices Databook Vol I 1984

AD are heavily committed to op-amp technology and this is a compact but comprehensive guide to specifying the devices. The data book also contains useful applications notes on interference shielding and guarding, grounding and decoupling.

On passive components:

Technical papers published between 1981 and 1988, Wima, 1989

A collection of informative articles on the technology of plastic film capacitors.

Introduction to Quartz Crystals, in ECM Electronics Ltd components catalogue, 1989

All the information you need on specifying and designing circuits for crystals.

Soft Ferrites: General Information, in Neosid Magnetic Components data book, 1985

A clear introduction to the properties and manufacture of ferrite components.

On power supplies:

Linear/Switchmode Voltage Regulator Manual, Motorola, 1989; **Voltage regulator Handbook,** National Semiconductor, 1982

Both these books, besides offering data sheets, contain chapters on aspects of power supply and regulator design.

Unitrode Applications Handbook, Unitrode, 1985

Unitrode specialise in power conversion semiconductors and this handbook contains a comprehensive selection of application notes covering switching supply design.

On printed circuits:

An Introduction to Printed Circuit Board Technology, J A Scarlett, Electrochemical Publications Ltd 1984

Introduces all aspects of standard, PTH and multilayer board manufacture. Also covers design rules.

BS6221 : Part 3: 1984, **Guide for the design and use of printed wiring boards,** British Standards Institution.

Exactly as the title describes it. Other parts of BS6221 cover assembly and repair of pcbs, as well as specification methods for each type (equivalent to IEC326)

Printed Circuits Handbook, ed. Clyde F Coombs, 3rd edition, McGraw Hill 1988

Well-established reference handbook.

On reliability:

Electronic equipment reliability, J C Cluley, Macmillan, 1981 (2nd Edition)

Useful overview of statistical basis, prediction, component failure rate and design

On transient suppression:

Transient voltage suppression manual, General Electric, 5th Edition, 1986

GE are major suppliers of varistor transient suppressors. This manual offers an overview of voltage transients and how to suppress them with varistors.

Application notes on Transient Suppression, Semitron

Similar to the above but concentrating on avalanche-type transient absorbers.

Index